NEAR MISS REPORTING AS A SAFETY TOOL

NEAR MISS REPORTING
AS A
SAFETY TOOL

Edited
by

T.W. van der Schaaf, D.A. Lucas and A.R. Hale

Butterworth-Heinemann Ltd
Linacre House, Jordan Hill, Oxford OX2 8DP

 PART OF REED INTERNATIONAL BOOKS

OXFORD   LONDON   BOSTON
MUNICH   NEW DELHI   SINGAPORE   SYDNEY
TOKYO   TORONTO   WELLINGTON

First published 1991

British Library Cataloguing in Publication Data
A catalogue record for this book is
available from the British Library

Library of Congress Cataloguing in Publication Data
A catalogue record for this book is
available from the Library of Congress

ISBN 0 7506 1178 2

Printed in England by Clays Ltd, St Ives plc

Contents

Dr. Gerald R. Brown
address: University of British Columbia
               Department of Civil Engineering
               2324 Main Mall
               Vancouver
               Canada V6T 1W5
function: Associate Professor
experience: - human factors in traffic safety

Prof. Andrew R. Hale
address: Delft University of Technology
               Safety Science Group
               Kanaalweg 2B
               2628 EB Delft
               The Netherlands
function: Professor of General Safety Science
experience: - safety management systems
               - individual behaviour in the control of danger

Dr.Ir. Richard van der Horst
address: TNO Institute for Perception
               Kampweg 5
               3769 DE Soesterberg
               The Netherlands
function: Senior Researcher, Traffic Behaviour Research Group
experience: - observational techniques in traffic behaviour
               - human factors research in road traffic

Mr. Geoffrey Ives
address: Colenco Power Consulting Ltd.
               Parkstrasse 27
               CH-5401 Baden
               Switzerland
function: Principal Project Manager
experience: - safety and reliability studies in nuclear power plants
               - human factors and event reporting in nuclear power plants

Dr. Deborah A. Lucas
address: Human Reliability Associates Ltd.
               1 School House, Higher Lane
               Dalton, Wigan
               Lancashire WN8 7RP
               United Kingdom
function: Senior Consultant
experience: - role of human error in transportation, aerospace and nuclear
                 processing sectors
               - research on memory lapses

Dr. Michel Masson
address: Service de Psychologie du Travail et des Entreprises
              Faculté de Psychologie et des Sciences de l'Education
              5, Bld. du Rectorat, B32
              Université de Liège au Sart-Tilman
              B4000 Liège 1, Belgium
function: Research and teaching assistant
experience: - A.I. modelling of nuclear operator's cognitive behaviour
              - intelligent decision support systems
              - methodology of accident analysis in industrial environments

Prof. James T. Reason
address: University of Manchester
              Department of Psychology
              Manchester M13 9PL
              United Kingdom
function: Professor of Psychology
experience: - human error in everyday life
              - modelling of human error in industrial safety

Drs. Tjerk W. van der Schaaf
address: Eindhoven University of Technology
              Department of Industrial Engineering and Management Sciences
              P.O. Box 513
              5600 MB Eindhoven
              The Netherlands
function: Lecturer in Cognitive Ergonomics
experience: - industrial safety management
              - human-computer interaction in process control rooms
              - fault diagnosis support systems

Mr. Roger K. Taylor
address: British Rail Research
              London Road
              Derby DE2 8UP
              United Kingdom
function: Head of BRR Human Factors Team
experience: - research on the human contribution to railway safety

FOREWORD

This book is the result of a three day discussion meeting held in Eindhoven, the Netherlands, in September 1989. Its theme was "registration and analysis of near misses" and it brought together a dozen safety professionals, academics and consultants from Western-Europe and Canada. Almost all participants had practical experience with near miss and accident reporting schemes in either the process industry (both nuclear and chemical) or in transportation (road traffic, railways and air traffic control).

The participants were especially surprised by the high degree of comparability of both design issues and implementation factors, in spite of the great diversity in areas of application, types of reporting schemes, and organisational conditions. This led to the feeling that it would be valuable to collect and communicate these ideas and lessons to a wider audience of safety practitioners and researchers.

The editors and authors have tried to transform these ideas and lessons into a volume of proceedings that will be stimulating and informative mainly for safety professionals "in the field" but also for academics and consultants in the fast growing area of safety management.

The editors gratefully acknowledge the sponsorship of the Eindhoven discussion meeting by the Commission of the European Communities through a grant by its Medical and Public Health Research Programme (Concerted Action on Breakdown in Human Adaptation: 2. Performance Decrement).

We also thank the Eindhoven University of Technology and its Department of Industrial Engineering and Management Sciences for supporting the organising of the discussion meeting and the editing of these proceedings.

Finally Mrs. Van Baalen should be especially thanked for the skillful processing of the manuscript.

Eindhoven, July 1991

A.R. Hale
D.A. Lucas
T.W. van der Schaaf

Chapter 1

INTRODUCTION

Tjerk W. van der Schaaf
Department of Industrial Engineering and Management Sciences
Eindhoven University of Technology
Eindhoven, The Netherlands

The focus of this book is on near-miss reporting within different systems. Its coverage is somewhat broader than this, because it sets near misses in the context of accidents and errors, and some of its chapters overlap into these adjoining areas.

In this introductory chapter some of the different purposes of collecting and analysing near-miss data are first discussed. The context of safety systems management as a whole and its relationship with quality and reliability issues are briefly indicated. Subsequently near-miss situations are placed inbetween actual accidents on one hand and safety-related human behaviour on the other. This leads to a number of different possibilities to define what a "near miss" actually is. Different methods of collecting near-miss data are listed and references to case studies in later chapters are made. The chapter is concluded by giving a brief overview of the remainder of the book, aimed at guiding the selective reader to the most relevant chapters.

1. PURPOSES OF COLLECTING AND ANALYSING NEAR-MISS DATA

Three general classes of such purposes may be distinguished:
1. to gain a qualitative insight into how (small) failures or errors develop into near misses and sometimes into actual accidents;
2. to arrive at a statistically reliable quantitative insight into the occurrence of factors or combinations of factors giving rise to incidents;
3. to maintain a certain level of alertness to danger, especially when the rates of actual injuries and other accidents are already low within an organisation.

1.1. Qualitative insight

We may wish to get a better understanding of how serious accidents *might* occur in our plant or organisation; in that case we are aiming at *modelling* the "production" of such accidents within and by the system as a whole.

Because the role of *human* safety-related behaviour is of special interest we should acknowledge the fact that humans are not only interesting for the *errors* they make, but also for their *error-compensation behaviour* or *recovery*: unlike most other system elements humans often have the capability to correct their own (or others') earlier errors or even small system failures by a timely action. This human recovery aspect will be especially prominent in investigating *near*-miss situations.

From a safety management perspective a specific goal within this broader purpose is then to *identify likely factors* or system elements in the sequence of events leading to near misses which in their turn may be considered as *precursors to actual future accidents*. From such a qualitative analysis two ways emerge to reduce the likelihood of such actual accidents: *error-inducing factors* can be eliminated (or their potential impact weakened), and *recovery-promoting factors* can be strengthened (or even introduced) in the system.

Another type of qualitative insight could be the result of regularly discussing unusual or *unique near misses*: new or unexpected combinations of circumstances might trigger an "Aha" experience with safety staff members and other employees. This means that their set of *"possible accident scenarios"* is enriched. Not only does this help them in their general understanding of system safety but it should also keep them aware of the practical impossibility of listing all types of failure modes (for any system of some complexity) in advance: after all, in actual accidents very often the "impossible" or "unthinkable" happens (see also 1.3).

1.2. Quantitative insight

Near misses are usually estimated to occur one or two orders of magnitude more frequently than actual accidents. Many companies and especially those which have already reduced their accident rate to a relatively low level paradoxically cannot measure their "safety performance" in a reliable way: their database of actual accidents is far too small to distinguish random fluctuations from actual trends. Because of their abundance near misses may then be used for such statistical *monitoring* purposes.

Specific goals here might be to be able to point out to management which combinations of *factors* are *most frequently* encountered in the analyses. This would provide them with a rational decision rule of where to concentrate resources (of time, efforts and funds) in order to increase the safety level in a *more efficient* way.

Also the *effectiveness of actions* taken to improve safety could be monitored in a quantitative way: a preventive action aimed at a specific error type should result in a reduced frequency of such errors when analysing subsequent near misses; likewise, recovery promotion should be measurable by an increase in frequency of its occurrence later on.

1.3. Maintaining alertness

Persistent *motivation* to be aware of the dangers of one's workplace or of the system as a whole is crucial to any organisation's "Safety Culture", and therefore to the safety-related behaviour of all levels of its employees. An important advantage in this respect is that near-miss investigation provides a *preventive perspective* much more than accident investigation which is corrective by nature. This issue in fact comes down to the well-known contrast between proactive and reactive management. Specific goals within this last type of purpose are firstly to counteract the idea that problems have been solved because there have been no accidents recently: knowledge of reported near-miss situations may be used to stress the fact that these should be considered as *precursors* to real accidents (see also 1.1) in less "lucky" future circumstances.

Near misses also provide a wealth of *in-house* (and therefore *convincing*) *examples* of common errors and recoveries. These may be used in training programmes for future managers and operators and in specific safety promotion campaigns.

1.4. Relationship with Quality and Reliability

For the purpose of this book the "spin-offs" of near-miss reporting schemes for Quality and Reliability programmes, although important, must be considered as *positive by-products*. The main points of overlap between these areas are in fact rephrased versions of earlier statements in this chapter:
- the possibility of *measuring* system performance on a behavioural level in a quantitative way;
- raised *awareness* of system defects as a first step towards eliminating them;
- a "zero defects" *attitude* in terms of goal setting with respect to "acceptable risk": prevention (or timely recovery) is preferred over correction.

In the long run these points of overlap should lead to more cooperation between safety staff, reliability engineers and quality managers: they will eventually probably realize that many "root causes" of the problems in their respective areas of interest are very much alike, if not identical.

2.   ACCIDENTS VS. NEAR MISSES VS. BEHAVIOURAL ACTS

The general purposes mentioned in the first part of this chapter apply of course just as well to actual accidents and to behavioural acts (that is: errors and recoveries) related to safety. Although these general purposes *in principle* are common to these three levels of incident-investigation, *in practice* there are usually large differences, as shown in Figure 1.

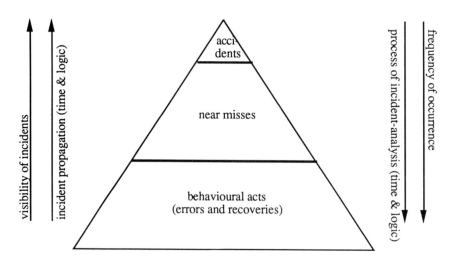

Figure 1.   A qualitative iceberg model of the relationships between accidents, near misses, and behavioural acts.

2.1.   A qualitative iceberg model

The familiar *"iceberg"* is shown here to indicate that near misses are "caught" in between actual, but rare, accidents on the top and an enormous number of errors and recoveries more to the bottom. *Incident propagation* is assumed to progress from the bottom to the top, which means that chances for early prevention of accidents decrease as you get closer to the top. The *order of incident analysis* is assumed to be top-down, but with different starting points in the iceberg depending on the type (or: level) of data that trigger the detection in the first place. It is also assumed that modern investigation techniques will always try to get as far to the bottom of the iceberg as possible and not stop at superficial descriptions of only the immediate events leading to an accident and its short-term consequences. Another vital assumption is that these three levels of the iceberg are directly related in the sense that they show largely overlapping sets of "root causes": a different starting level should *not* lead to an entirely different set of root causes being identified by the analysis, and should also then *not* lead to a fundamentally different set of suggested actions in order to tackle these.

The starting point of detecting and analysing incidents must therefore be determined by other dimensions, such as *frequency of occurrence* and the *"visibility"* of the incidents. Figure 1. shows the well-known phenomenon of very rare (in some companies even absent) accidents and an abundance of errors and recoveries. It also goes without saying that actual accidents have the highest visibility, but that day-to-day behavioural acts are easily overlooked, although their consequences in less forgiving environments might have been serious.

It is the main thesis of this book that for many companies and authorities near-misses may provide an optimum between highly visible (and detectable) but rare accidents, and very frequent but almost invisible behavioural acts, and that they are therefore worth collecting and analysing.

2.2. Defining Near Misses

At this point we may try to formulate a *suitable definition* of what constitutes a "near miss". Some authors may wish to restrict near misses to situations with a major human component, while others may prefer to include also purely technical hardware incidents or deviations. Another matter of debate is whether the avoidance of an otherwise much larger accident is also a "near" miss or if this label should be strictly reserved for incidents without any real (economic) effect.

We propose to use the following *"working definition"* for this moment (at the end of this book in Chapter 12 an evaluation of its suitability in the light of the case studies presented will be given):

"A near miss is any situation in which an ongoing sequence of events was prevented from developing further and hence preventing the occurrence of potentially serious (safety related) consequences."

. Stopping the incident sequence may have been brought about either by *"luck"* (that is: a random combination of circumstances) or by a purposeful action (*"recovery"*) which may have been planned beforehand (as in procedures or safety valves), or executed on an intuitive basis at the time of the incident.
. Consequences may include *damages* (material, production loss, environmental, etc.), *injuries*, or other negative effects, but they exclude mere psychological consequences, such as surprise, fright, etc., associated with experiencing such incidents and their effects.

As an illustration a few examples may be useful:
- Crane driver drops a load (dangerous occurrence)
  . it hits a person standing below (an accident),
  . no one is standing underneath at the time (near miss: chance factors),
  . a coworker pushes a person out of the way (near miss: human recovery),
  . the area under the crane is restricted (near miss: management control),
  . the crane design has an automatic stop device (near miss: technical safeguard).

- Air miss, that is two planes on a collision course
  . crash occurs (an accident),
  . one pilot sees other plane just in time to take evasive action (near miss: human recovery),
  . air traffic controller spots the conflict and orders new course (near miss: human recovery, but with different agent),

. warning device in cockpit detects collision course (technical safeguard requiring some human action or decision),
. one plane just happens to change altitude before time of impact (chance factors).

3.   METHODS FOR COLLECTING NEAR-MISS DATA

Near misses as defined in the previous paragraph may be collected by way of several possible techniques. They may be *reported* by the persons "experiencing" the near miss on either a voluntary or mandatory basis. They may however also, because of their "visibility", be *observed* by registration equipment or human observers. Finally they may be *generated* in experimental conditions, usually by means of complex *simulation* facilities.

3.1. Reporting-based methods

These methods expect the employees themselves to report on such incidents as part of their job; usually references are made towards preventing accidents happening to less lucky colleagues in the future, or it may be required or expected in the course of some Total Quality Programme. Examples of *voluntary reporting* are given in Chapters 5 and 6, both schemes from the process industry. *Mandatory reporting* seems to be more prevalent in some sectors of transportation, as shown by the fact that both pilots and air traffic controllers are legally required to report air misses in most countries.

3.2. Observation-based methods

Outsiders with respect to the chain of events leading to a near miss may also be used to detect such incidents. A clear example of *automatic registration* is described in Chapter 8 on the passing of railway signals at danger. Also the famous "black boxes" in aircraft may be put into this category, where they are not only investigated after an accident, but on a routine basis after each flight to detect excursions outside planned flight parameters. In systems where near misses may be expected to occur predictably under certain system conditions (like starting up a plant) or at regular intervals (like rush hours in a congested city) *human observers* may be trained to detect them. Traffic conflicts (no crashes or injuries) are examples handled in Chapters 9 and 10, and in one of the cases reported in Chapter 5 engineers were on stand-by to observe operator actions under critical conditions. In such observations usually only a *sample* is taken from the set of all locations and all opportunities; otherwise it would soon demand astronomical resources.

3.3. Simulation-based methods

A more experimental approach is given in Chapter 4 which deals with simulation facilities. These may be used to *generate* errors, recoveries, near misses and "accidents" on the basis of suitable scenarios; because the conditions are under the control of the experimenter very efficient data collection is possible, but the question is always whether these data are *valid* and therefore generalisable to the real world.

Another way of using simulation facilities is for *modelling* purposes: the effects of time-stress on fault diagnosis for instance could be modelled in this way, and frequent errors and recoveries could then be used to arrive at suggestions for decision support and interface design.

3.4. Selecting a particular method

It is very difficult at this stage to advise on one (or more) of the above methods in a particular situation. The main question to be answered first is *which purpose(s)* should have priority (see paragraph 1 of this chapter). Even so, at least *four other aspects* must be taken into consideration:
- *level and visibility* of the "dangers" involved; highly visible high-consequence situations could favour voluntary reporting. Dangers which are less obvious to the reporting employees suggest the use of automatic recording.
- *amount and depth of data required*: observation-based methods may "produce" many more instances of near misses, but with less depth than reporting one's own (partly invisible) diagnostic misinterpretations for instance. A combination of these two methods is shown in Chapter 8.
- *phase of the (production-)system*: in the design phase a simulation/modelling approach would probably be more fruitful than when production has already been started and changes in the hardware have become very expensive.
- *acceptability to the employees*: automatic recording can give rise to concerns among employees who fear a "Big Brother" regime spying on them. Voluntary recording will only work with high personnel motivation (see Chapters 5, 6, 11 and 12).

4.   A BRIEF OVERVIEW OF THE BOOK

In this final section of the Introductory Chapter a brief overview of the remainder of the book will be presented.

4.1. General principles

The following chapters will deal with general principles of systems for managing industrial safety but with an emphasis on the role of near-miss reporting schemes.
. *Reason* (Chapter 2) focusses on the importance of bad management decisions in the design phase which may cause severe problems for safety management much later. These build "latent pathogens" into the system, ready to surface and open up "windows of opportunities" for fatal combinations of failing system elements.
. *Van der Schaaf* (Chapter 3) presents a framework to describe or design complete systems for "managing" near-misses; to get them reported, described, analysed and interpreted into suggestions for actions. The seven

modules or steps which together make up this *Near Miss Management System* (NMMS) framework are also used as headings to describe some of the cases and to organise their conclusions.

. Finally *Lucas* (Chapter 11) complements the design-oriented NMMS framework by discussing the organisational and management aspects which are crucial in embedding such a NMMS within the culture and tradition of a company or authority. These aspects also receive much attention in the preceding chapters describing the case studies, and will be reviewed again in the last chapter where the conclusions are drawn and some initial lessons are formulated.

4.2. Case studies

The case studies are taken from nuclear power plants (Ives in Chapter 5, and Masson in Chapter 4), chemical process plants (Van der Schaaf in Chapter 6, and Hale -in part- in Chapter 7), and from several areas of transportation, namely railways (Taylor and Lucas in Chapter 8) and road traffic (Van der Horst in Chapter 9, and Brown in Chapter 10). They describe not only "historical" cases (e.g. Ives, Taylor and Lucas) but also projects which are still under development (e.g. Hale, and Van der Schaaf). Another dimension to characterize these cases is whether they describe successes (e.g. Taylor and Lucas) or spectacular failures (e.g. Ives, Hale -in part-).

Chapter 2

TOO LITTLE AND TOO LATE: A COMMENTARY ON ACCIDENT AND INCIDENT REPORTING SYSTEMS

James Reason
Department of Psychology
University of Manchester
Manchester, U.K.

ABSTRACT

Safety, like health, is a difficult notion to pin down and an even harder one to measure. By comparison, unsafe states (like diseases) are all too clearly signalled by fatalities, injuries, physical damage and financial losses. Each of these negative aspects readily translates into numbers of one kind or another. So should we not settle, as many organisations have, for assessing the relative safety of their various activities by the number and severity (actual or potential) of the incidents and accidents they sustain over a given period?

This chapter proposes that while incident and accident reporting systems are a necessary part of any safety information system, they are, by themselves, insufficient to support effective safety management. The information they provide is both too little and too late for this longer-term purpose. In order to promote proactive accident prevention rather than reactive "local repairs", it is necessary to monitor an organisation's "vital signs" on a regular basis. Only these systemic factors lie within the organisation's direct sphere of control. Moreover, it is in the fallible decisions taken within these organisational and managerial sectors that most accidents to complex, well-defended systems have their principal origins (Reason, 1990a). In short, we are now in the age of the "organisational" accidents.

1. TRACING THE HISTORY OF "ORGANISATIONAL" ACCIDENTS

Traditionally, incident and accident investigations begin with the occurrence of some catastrophic event and then work backwards trying to find answers to three main questions: How did the accident happen? Who was responsible? By what means could a recurrence be avoided? Accidents almost always arise from the adverse conjunction of several causal chains. In theory, each one could be traced back as far as records will allow. Cecil Woodham-Smith (1953), for example, in seeking the reasons for the disastrous Charge of the

Light Brigade, found it necessary to begin her story two centuries earlier with the military legacy of Cromwell's Major-Generals. But accident investigators are not historians. A mainly technical orientation combined with limited investigative resources impose strict stop rules on their enquiries. Although these rules are rarely made explicit, accident investigations generally end when immediate answers are found to the cause, responsibility and recurrence questions. In the past, this has usually meant describing the events just prior to the accident, identifying those people or equipment items whose failures were directly responsible for the system breakdown and specifying a list of local repairs that, had they been in place, would have thwarted the accident sequence.

To a large extent, the scope of an accident investigation is determined by the kinds of preventative measures required. When the unlucky William Huskisson was fatally injured by the Rocket at the opening of the Liverpool and Manchester Railway in 1830, the counter-measures were fairly obvious: introduce rules and physical barriers to keep vulnerable flesh out of the way of heavy metal. Later, as the railway network was extended, it was necessary to develop more elaborate defences: signals, improved brakes, the absolute blocking system, and so on. These engineering and operational measures greatly improved the safety of rail travel, but they did not eliminate serious accidents.

In 1929, at the approaches to Paragon Station, Hull, two signalmen independently committed just those errors necessary to create a two-second chink in a state-of-the-art safety system of interlocking points, bringing an incoming train into collision with a stationary one (Rolt, 1978). Accidents like this marked the transition point between the engineering age and the human error age. An early concern of this new age was the possibility of identifying accident prone individuals.

The idea that certain personal characteristics might dispose some people to have more than their fair share of accidents was born in the 1930s when engineers were finding it increasingly difficult to reduce accidents by technical safeguards. For a short while, it seemed as though the answer to the residual accident problem was at hand: identify these accident-disposing traits and then establish selection procedures to eliminate their possessors from the workforce. Unfortunately, the accident-proneness 'club' turned out to have a membership that was both widely assorted and continually changing. Subsequently, human factors specialists devoted themselves mainly to mitigating the physical and psychological insults created by design engineers for whom the human-machine interface was often an afterthought, an area into which the bare control and display necessities were located as and where space could be found for them.

These two ages of safety concern, the technical and the human, are clearly recapitulated within the short span of nuclear power generation. In the 1960s and early 1970s, safety measures were directed primarily at minimising the consequences of technical failures: loss of coolant accidents (LOCAs), steam generator tube ruptures and the like. Much of the reliability debate centred upon the effectiveness of engineered safety devices such as the emergency core cooling system. This has been described as the time of the "LOCA mindset" (Ward, 1988).

It is difficult to identify the precise beginning of the nuclear industry's human error era, but the Brown's Ferry candle is as good a marker as any. In 1975, two maintenance technicians, tracking down an electrical fault, used a candle to light their way and started a fire which knocked out all of the plant's automatic safety features. This industry's human error concerns passed through two distinct phases. They began with an almost exclusive interest in execution failures (slips and lapses), then, after Three Mile Island in 1979, the emphasis broadened to embrace diagnostic errors and the selection of inappropriate recovery strategies (mistakes).

For many within the nuclear industry, the principal lessons of Three Mile Island were limited to the costly demonstration that operators can make serious errors of judgement in emergency conditions (see Layfield, 1987). Subsequent remedial measures (better training, improved control room displays, the provision of decision support systems) focused mainly upon minimising the recurrence of these mistakes. But for others, however, this extensively investigated incident marked the start of a third safety era: the age of the organisational accident, or the socio-technical era.

The central importance of socio-technical factors in accident causation has been amply reinforced by the last three major accident inquiries to be held in the UK (relating to the accidents at Zeebrugge, King's Cross and Clapham Junction) which have seen a dramatic change of emphasis. While all three inquiries involved a close investigation of the active failures committed by those at the 'sharp end', the major part of each was taken up with tracing the lines of responsibility back from the 'sharp end' to the managerial and organisational spheres. In these accidents, as at Three Mile Island, Bhopal, Chernobyl and Piper Alpha, it soon became apparent that the principal origins of disaster lay in the fallible decisions of those within the upper reaches of the organisation. In many cases, these decisions had been taken several years before the accident actually occurred.

A close study of these recent 'organisational' disasters has prompted accident researchers and investigators to widen both the scope and the retrospective timescale of their enquiries. It is now apparent that we need to distinguish two ways in which human beings can contribute to technological disasters:

Active failures
These are errors and violations having an immediately adverse effect upon the system's integrity. They are committed by those at the 'sharp end', and usually involve the direct, though not necessarily deliberate, circumvention of the various layers of the system's defences.

Latent failures
These are decisions or actions, the damaging consequences of which may lie dormant for a long time, only becoming evident when they combine with local triggering factors (that is, active failures, technical faults,

atypical system conditions, etc.) to break through the system's defences. Their defining characteristic is that they were present within the system well before the onset of a recognisable accident sequence. They are most likely to be spawned by those whose activities are removed in both time and space from the direct human-machine interface: designers, system builders, high-level decision makers, regulators, managers and maintenance staff.

Perhaps the greatest threat now facing well-defended hazardous technologies stems not so much from the breakdown of a major component or from isolated operator errors as from the insidious accumulation of latent human failures within the organisation.

2.   THE RESIDENT PATHOGEN METAPHOR

It has been suggested (Reason, 1988, 1990b) that latent failures in technical systems are analogous to resident pathogens in the human body which combine with local triggering factors (i.e., life stresses, toxic chemicals and the like) to overcome the immune system and produce disease. Like cancers and cardio-vascular disorders, accidents in defended systems do not arise from single causes. They occur because of the adverse conjunction of several factors, each one necessary but none sufficient to breach the defences. As in the case of the human body, all technical systems will have some pathogens lying dormant within them.

This view leads to a number of very general assertions about accident causation.

(a) The likelihood of an accident is a function of the number of pathogens within the system. The more abundant they are, the greater is the probability that some of these pathogens will encounter just that combination of local triggers necessary to complete a latent accident sequence. Note that this view demands quite a different calculus than that employed in conventional probabilistic risk assessment.

(b) The more complex and opaque the system, the more pathogens it will contain.

(c) Simpler, less well-defended systems need fewer pathogens to bring about an accident.

(d) The higher a person's position within the decision-making structure of the organisation, the greater is his or her potential for spawning pathogens.

(e) Local triggers are hard to anticipate. Who, for example, could have anticipated that the assistant bosun of the Herald of Free Enterprise would be asleep at the time he was required to close the bow doors, or that the chief officer would have mistaken someone else walking towards the bow doors on the car deck just before sailing time as the assistant bosun? But it would have been possible to establish in advance that the ship was undermanned and poorly tasked.

(f) The key assumption, then, is that resident pathogens can be identified pro-actively, given adequate access and system knowledge.

(g) It also follows that efforts directed at identifying and neutralising patho-gens (latent failures) are likely to have more safety benefits than those directed at minimising active failures. Another way of expressing this is to state that it is more effective to remove organisational failure types (the "parents") than individual failure tokens (their "offspring"). This dis-tinction between types and tokens is discussed further below.

3.   ORGANISATIONAL FAILURE TYPES AND INDIVIDUAL FAILURE TOKENS

Types and tokens are both classes of human failure. They are distinguished in two ways: by their degree of specificity and by their points of reference within the productive element framework. Types are general classes of organisational and managerial failures. Tokens are more specific failures relating to individuals at the human-system interface.

Types break down into two sub-categories, each relating to a productive element. Source types are associated with fallible decisions at the strategic apex of the organisation (Mintzberg, 1979). As the term suggests, they are seen as the principal systemic source of pathogens and are discussed further below. Function types relate to the line management element. This is where fallible strategic decisions are translated into functional forms and then distributed throughout the organisation along departmental pathways.

Tokens are similarly divided into two sub-categories: condition and unsafe act tokens. Condition tokens comprise the psychological or situational states conducive to the commission of unsafe acts. They may be further sub-divided into three groups: information-processing factors (relating to atten-tion, memory and knowledge); situational factors (having to do with such things as the ergonomic quality of the human-system interface, workload, distractors and the like); and social and motivational factors (attitudes and group norms).

Unsafe act tokens subdivide into three groups: slips and lapses (i.e., deviations of action from intention); mistakes (i.e., deviations of planned actions from some adequate path to the desired goal); and violations (deliberate infringements of safe working practice). It was violations rather than errors that contributed to the Chernobyl disaster (Reason, 1987).

The principal advantages of the type-token distinction are, firstly, that it allows us to specify more precisely the way in which fallible strategic de-cisions translate into unsafe acts; secondly, it directs us away from the dangers of 'tokenism': the tendency to treat human failures at a local level rather than within the global context of the organisation.

4.   A MODEL OF ACCIDENT CAUSATION

Source types, as mentioned earlier, are the principal origin of resident pathogens. In essence, they relate to the way in which top managers choose to allocate finite resources between production and safety goals (see Reason, 1990b). These decisions and their underlying attitudes create a system-wide safety culture. Failure types at this level are then amplified and distributed throughout the organisation by the activities of the function specialists (operations, maintenance, training, design, etc.). There is thus a few-to-many mapping between source types and function types. Figure 1 summarises the nature of these mappings between the causal elements of the model.

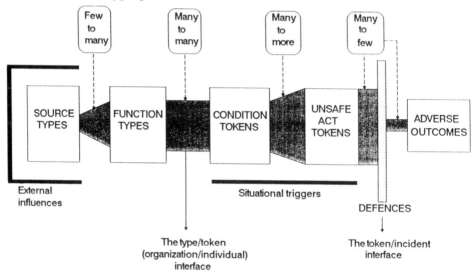

Figure 1.   Showing the nature of the onward mappings between the type-token elements of the accident causation model.

From a theoretical point of view, the most critical onward mappings lie at the organisation/individual interface, where function types are translated into condition tokens: the situational and psychological precursors of unsafe acts. These translations are of a many-to-many kind. That is, any one function type (i.e. poor tasking, deficient training, inadequate procedures, error-enforcing instructions, etc.) can give rise to many condition tokens (i.e., high workload, inappropriate perception of hazards, ignorance of the system, etc.). Similarly, any one condition token can be the product of many different function types.

This multiple determination of condition tokens means that we cannot simply use the observed presence of a particular kind of condition token to infer the existence of specific failure types upstream in the organisation. The same problem exists at the next set of linkages: condition tokens to unsafe act tokens. Not only is each unsafe act the product of many possible type "parents", it is also likely to have been "midwifed" by a unique set of triggering conditions.

The complexity of these interactions also means that backtracking from tokens is insufficient to identify the nature of the originating failure types. To achieve this, it is necessary to employ a variety of proactive indicators.

5.   THE TWO FACES OF SAFETY

Safety has two faces. There is the harsh face revealed by accidents, incidents and near misses. These reflect an organisation's periodic vulnerability to adverse combinations of operating hazards, situational factors and human or technical failures. There is also a positive but largely concealed face of safety. This relates to an organisation's potential for resisting future adverse conjunctions of hazard and failure. Such intrinsic resistance has two components: (a) the scope and effectiveness of the system's defences and (b) its basic "safety health" as derived from the essential quality of its various functions and activities.

Some organisations assert that all accidents are avoidable. And so they may appear -- with hindsight. When reading accident reports, it is hard to resist the "if only" game: "If only X had been more careful and Y hadn't done this, there wouldn't have been accident." But X and Y almost certainly did not realise that they were actors in a much larger accident scenario. They were probably doing the things they had often done before, or were committing previously inconsequential errors and violations (see Wagenaar & Groeneweg, 1987). Accidents, by definition, are unwanted and unexpected events. Just as we do not will them to happen, so we cannot simply will their non-occurrence.

"Early and provident fear," wrote Edmund Burke, "is the mother of safety." If that were the whole story, if intelligent unease, care and prudent precautions were sufficient to guarantee immunity from accidents, then accident reporting systems would indeed provide an accurate picture of organisational safety states. If accidents were solely the result of an organisation's negligence or indifference, then they should also serve to reveal the nature and the extent of these shortcomings. But that is far from being the case. Murphy, Sod, continuing hazards, ineradicable human fallibility and unhappy chance also have large parts to play. In short, accidents are caused by both deterministic and stochastic factors.

The more prevalent tendency in the past, however, was to emphasise the stochastic factors at the expense of more deterministic ones. In the early nineteenth century, Poisson, the French mathematician, recorded the number of horsekicks sustained by Prussian cavalrymen over a given period. He noted that the largest proportion suffered no kicks and that diminishingly smaller proportions received increasingly larger numbers of kicks. The Poisson distribution is now used as a theoretical model for determining the chance probability of an accident among a group of individuals sharing equal exposure to hazard.

But the Poisson distribution takes no account of the possibility that the large group of accident-free people might possess differing degrees of resistance to the hazard in question, and that these variations may have little or nothing to do with chance. For example, some of the unkicked cavalrymen may have owed their immunity not to luck but to a good relationship with their mounts, others to the extreme caution they exercised when operating in the 'kicking zone', and so on. In other words, it is possible to envisage a scale that discriminates among the kick-free cavalrymen in terms of their relative safety.

An individual's position along this safe-unsafe dimension will be determined by many different factors. In some cases, it will simply be a question of good luck. But in others, it will be due to the success of deliberate countermeasures. The most resistant cavalrymen will be those who employ the largest number of preventatives (kindness, caution, etc.) in the most sustained and effective way. Conversely, it is likely that among the most kick-prone will be those who were both unkind and careless as well as unlucky. Chance, of course, will play its part, but it will not be the whole story; deliberate actions as well as luck determine the safety state.

6.   INTRODUCING THE "SAFETY SPACE"

Let us now apply these arguments to organisations rather than individuals, or to different activities within some very large company. It is possible to imagine that organisations engaged in potentially hazardous operations might, in any one period, occupy different positions along some relatively safe to relatively unsafe dimension. This position will be a function of many deterministic (as opposed to stochastic) factors.

Expressing this in a more visual way, it can be suggested that these degrees of organisational safety or unsafety can be represented by different locations within a notional "safety space" (see Figure 2). This space is cigar-shaped to indicate that the majority of organisations are likely to occupy the middle ground, with numbers diminishing as one approaches either the relatively safe or relatively unsafe extremes. These positions are assumed to reflect the influence of deliberate (non-stochastic) safety measures; that is, they reflect the success or otherwise of an organisation's current safety programme. It is presumed that the safety programme acts directly upon an organisation's intrinsic resistance or vulnerability to the hazards associated with its particular modes of operation.

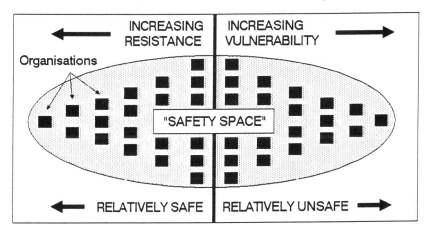

Figure 2.   Showing how 40 hypothetical organisations might be distributed within the "safety space" at any one time. Their positions are determined by the quality or otherwise of their current safety measures.

Figure 3 represents the presumed relationship between an organisation's position on the safe-unsafe dimension and the likely influence of stochastic factors. The basic assumption is that resistance or vulnerability to unhappy or happy chance increases as one moves towards one or the other extreme of the safety 'space'. It is those organisations occupying the middle ground that are most likely to be affected by these random (or at least unpredictable and ungovernable) factors one way or the other.

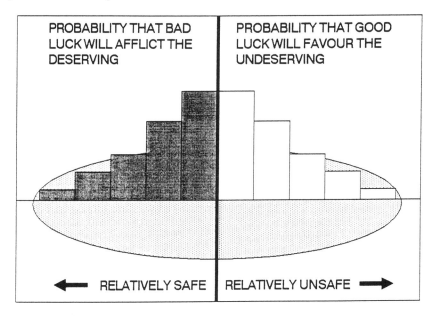

Figure 3.   Showing how an organisation's position within the "safety space" interacts with stochastic factors. The influence of these factors (for good or ill) diminishes as an organisation approaches either extreme of the "safety space".

   This means that while intrinsically safe organisations can, to a large extent, "make their own luck", they are not immune to accidents. No organisation, no matter how committed or competent, can eliminate human fallibility, nor can it entirely remove the hazards associated with its operations. The same argument applies in reverse to the unsafe extreme: even the undeserving can have runs of good luck. In short, there is no such thing as absolute safety or absolute unsafety. Although this view predicts a positive relationship between accidents and degrees of unsafety, the correlation coefficient is likely to be much less than unity. But, notwithstanding the attenuation of this relationship due to chance interventions, the "safer" organisations are likely to fare better in any particular sampling period. While not immune from accidents, their planned counter-measures will offer greater protection against damaging concatenations of chance and hazard.

7.   NAVIGATING THE "SAFETY SPACE"

Effective safety management means actively navigating the safety space in order to approach and then remain within the safety zone. To do this, safety managers must appreciate the nature of their various navigational aids and what kind of information is provided by each of them. They must also understand what are the factors that can drive their organisation toward the safety zone. These issues are summarised in Figure 4.

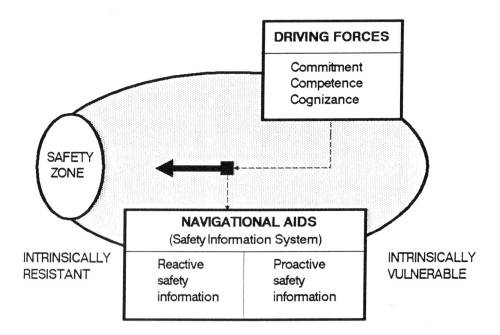

Figure 4.   Summarising the principal factors involved in navigating the "safety space": (a) the driving forces of commitment, competence and cognizance, and (b) the navigational aids supplied by the organisation's safety information system.

The driving forces

The "driving forces" are the principal determinants of an organisation's current and future positions within the safety space. They equate to the source types, referred to earlier. Three core factors determine the nature and extent of the source types at the strategic apex of a system: commitment to safety goals, competence to achieve safety goals and cognizance of the nature of the dangers facing the system.

Commitment
This factor has two principal components: the one concerned with motivation, the other with resources. The motivational factor relates to whether an organisation seeks to be a market leader in safety within its particular sphere of operations, or whether it is simply content to keep one step ahead of the regulators (see Westrum, 1988). The second issue concerns the resources allocated to the attainment of safety goals. This is not just a question of money, it also has to do with the calibre and status of the people assigned to manage system safety.

Competence
Commitment by itself is not enough; the organisation must also possess the technical competence to achieve its safety goals. The competence factor may also be broken down into two components: one relating to the quality of the organisation's safety information system, the other concerned with how it responds to safety-related information.

Cognizance
No amount of commitment or even competence will suffice unless the organisation has a proper awareness of the potential hazards endangering its operations. In essence, cognizance boils down to two questions. First, does the organisation understand the true nature of the "safety war"? Second, can it learn the right lessons from its own safety-related experiences?

8.   SAFETY INFORMATION SYSTEM CHANNELS

In order to direct their organisation to the intrinsically resistant end of the safety space, managers need to take regular samplings of the organisation's "vital signs" or proactive safety state indicators. Figure 5 shows five possible information channels (which together comprise the safety information system) derived from the model of accident causation described earlier. Their characteristics are discussed below.

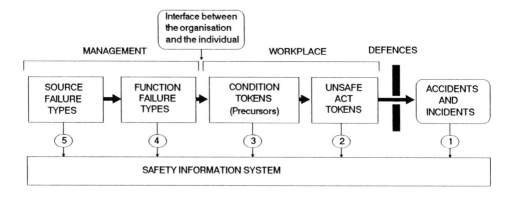

Figure 5.   The basic elements of a safety information system as they relate to the type-token model of accident causation.

8.1. Channel 1: Accident and incident reporting systems

This is the most widely used channel, yet it is the least useful for effective safety management. The information it provides is both too "noisy" (for the reasons discussed at length earlier) and too late in the causal sequence. But it is not without value. Each accident or incident reveals a specific "pathway" of accident opportunity through the system's defences. Taken individually, accident and incident reports tend to lead to local solutions: the modification of procedures, engineering "retrofixes", the retraining or disciplining of personnel. Over time, these can gradually yield improved procedures and strengthened defences. But there are more cost-effective ways of achieving this. When a large number of accidents or incidents are examined collectively, it is sometimes possible to identify recurrent patterns within the accident-producing circumstances. More commonly, however, the most conspicuous result of such analyses is the remarkable absence of significant patterns or trends. In many large organisations, the main use for Channel 1 data is to produce league tables of unsafety by which to identify "the good, the bad and the ugly" among their various activities or operating companies.

8.2. Channel 2: Unsafe act and near miss auditing

On the face of it, the information supplied by this channel should be more valuable than that provided by Channel 1. Unsafe acts (errors and violations) and near misses are the stuff of which accidents are made. The numerous "iceberg" theories of accident causation presume various (widely differing) ratios between unsafe acts, near misses, lost time injuries and fatal accidents. But there are serious measurement problems. The actual numbers of unsafe acts committed are almost impossible to determine. What is certain is that these numbers are extremely large.

   A very conservative estimate by an experienced unsafe act observer

(Groeneweg, 1991) was 7 unsafe acts per worker per hour. In a company employing 6000 people, this would yield approximately 6 million unsafe acts per year. Measuring, let alone eliminating, these 6 million unsafe acts would be an impossible task.

Of much greater value than the raw numbers is information relating to the nature and variety of unsafe acts associated with particular types of activity. Different kinds of unsafe act demand different modes of remediation. Errors are the result of information-processing problems and require either improved workplace design or better training, or both. Violations, on the other hand, have their origins in motivational, attitudinal and group factors and need to be tackled by counter-measures aimed at the heart rather than the head.

Qualitative information of this nature can be obtained through a variety of auditing techniques. Du Pont's Unsafe Act Auditing, for example, involves frequent observations (1-2 hours per week per worker) by local supervisors of individual working practices. Afterwards, the things that did or could go wrong and the ways in which the job could be done more safely are discussed with the workforce. Here, then, the feedback loop runs only as far as the supervisory level, and thus has limited value as a safety management tool. But it does have two very useful functions. It raises individual safety consciousness and, at the same time, communicates top-level concern with safety.

But even if it could yield valid and timely information (which is highly unlikely), Channel 2 would only be of limited value for those concerned with the more strategic issues of safety management. Unsafe acts, along with LTIs and accidents, are tokens not types. They are like the dragon's teeth in the fairy tale: eliminate one and another will appear in its place. Errors and violations are part of the human condition. They can be moderated and guarded against, but they can never be eliminated altogether.

8.3. Channel 3: The precursors of unsafe acts

This channel deals with the environmental and psychological precursors of unsafe acts: poor workplace design, high workload, unsociable hours, inadequate training, poor perception of hazards, and the like. They are one stage upstream from unsafe acts in the accident causation model and hence far fewer in number, but they are still tokens; that is, they remain only expressions of higher level failure types.

In principle, information relating to these condition tokens would be of considerable use at the supervisory and line management levels. In practice, however, it is extremely difficult to collect and process such data proactively. Given the benefits of hindsight, it is easy enough to recognise the contribution of these factors to individual accidents. But it is quite another matter to foresee their contribution to future accidents. Some of these precursors are a natural, even necessary, part of normal operations, yet others are unchangeable within current working constraints and still others may play no significant role in any future accident scenario.

The difficulty with both condition and unsafe act tokens is that, by themselves, they are rarely sufficient to cause accidents, particularly in complex, well-defended systems. Their adverse consequences arise through multiple and often unforeseeable conjunctions with local triggering conditions. In short, they are subject to the vagaries of chance and hazard. As such, they are neither particularly informative nor readily remediable.

8.4. Channel 4: Failure type indicators

Only at the level of the function failure types is it possible to start gaining direct access to the "vital signs" that reveal the intrinsic safety health of an organisation. And it is mainly within this organisational (as opposed to individual) domain that safety managers can begin to exert effective and targeted control over the system.

If each function (department) of an organisation were wholly independent of all the others, we could only assess a company's overall safety "health" by measuring all of its elements individually. Alternatively, if all these elements were closely related to one another, then the state of any one of them would provide a global picture. It is likely that the reality lies somewhere between these two extremes.

One approach is to focus upon a limited set of general failure types (GFTs) whose indicators are derived empirically from close observation of various organisational activities. This is the method adopted by a research group from the Universities of Leiden and Manchester in a three-year project sponsored by Shell Internationale Petroleum Maatschappij in The Hague. This work has led to the identification of 11 GFTs and their associated activity-specific indicators. A hypothetical Failure State Profile based upon these 11 GFTs is shown in Figure 6. General failure types are abstractions. They can only be assessed through their concrete and tangible manifestations. Each GFT reveals itself in a variety of ways, depending upon the nature of the activity being audited. The main aim of the Shell research has been to develop these "hands on" indicators for a wide range of exploration and production activities (i.e. seismic, drilling, North Sea gas platforms, transport, maintenance, construction, etc.). Typically, there are something like 130 interchangeable checklist items for each activity-specific Failure State Profile (FSP). In addition, FSPs can be derived from global ratings of each of the 11 GFTs made by experienced supervisors or managers. Positive and significant correlations have been obtained (a) between the checklist and global rating methods of generating FSPs, and (b) between FSPs and judged contributions to actual accidents (of each of the 11 GFTs) for a given activity (North Sea gas platforms), where both sets of assessments were made by independent observers. These findings suggest that FSPs are both reliable and valid indices of the intrinsic safety state of a given activity or function.

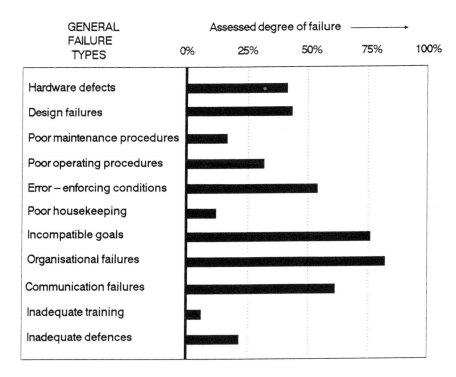

Figure 6. A hypothetical Failure State Profile based upon assessments of the 11 General Failure Types.

8.5. Channel 5: Stylistic or cultural indicators

Whereas Channel 4 information provides regular and frequent checks upon a system's "vital organs", Channel 5 data yield the basis for longer-term assessments of the organisation's global safety style. More specifically, this channel relates to the top-level commitment, competence and cognizance factors discussed earlier. Since these "cultural" features are likely to change rather slowly, the sampling rate need only be at something like yearly intervals.

At present, there is no standardised form for these stylistic assessments. However, the following 7-point rating scale summarises some of the issues that might be considered.

1. *Pathological*
   Safety practices at the barest industry minimum. No top-level commitment to the pursuit of safety goals.

2. *Incipient-reactive*
   Keeping just one step ahead of the regulators, but showing some signs of concern about accident trends.

3. Worried-reactive

Beginning to get seriously worried about continuing runs of incidents or accidents.

4. Repair-routine

Reasonable sensitivity to past events and possible future ones. Safety data (Channel 1) collected and analysed, but problems dealt with only by local repair actions.

5. Conservative-calculative

Possess a wide range of auditing techniques and workplace safety measures but still highly "technocratic" in their remedial measures. The organisation remains firmly locked into the technical and human error safety eras.

6. Incipient-proactive

Characterised by an early awareness that "engineering fixes", selection, training and motivating are not enough. Actively searching for better solutions. Beginning to acknowledge the importance of organisational and managerial factors.

7. Generative-proactive

Many proactive measures in place (i.e. Channels 4 and 5). Organisational safety measures under constant review. Top-level commitment to improving safety culture. A range of diagnostic and remedial measures continuously being reviewed and implemented. A marked absence of complacency.

Good readings on Channel 5 are the prerequisite of an effective safety programme. Both source failure types and intrinsic safety health begin at the top.

9.   CONCLUSION: MANAGING THE MANAGEABLE

The key to effective safety management lies in appreciating what is controllable and what is not. Many organisations seek to exercise direct control over the reduction of incidents, lost time injuries (LTIs) and accidents. They apply the same basic model to the safety arena as they do to production. Instead of setting production goals, they set safety goals in terms of some reduction in LTIs and fatalities. Instead of implementing the various means of production, they apply a variety of often off-the-shelf safety measures.

In the case of production, the feedback channel from products to production goals is relatively "noisefree". Most of the factors which influence the quantity and quality of products lie within the sphere of influence of the organisational managers. But the same is certainly not true of accidents, LTIs and the like. These result from a complex interaction between two distinct sets of causal factors: (a) latent failures or resident pathogens within the organisation, and (b) local triggering events, which are very much the property of either the human condition (human fallibility) or the external environment (chance and hazard). As a consequence, the feedback provided by conventional incident and accident reporting constitutes a very "noisy" channel indeed, having only a tenuous connection to the efficacy of the safety measures employed by the organisation. Organisational failings are by no means the only causes of accidents. There is a large stochastic element involved. Thus, "healthy" organisations can still have bad accidents, while many "sick" organisations may escape retribution for long periods of time. Chance conjunctions and ever-present hazards can cause accidents despite the safety measures implemented by the organisation.

Accidents, by their nature, are not directly controllable by an organisation. So much of their causal variance lies outside the organisation's sphere of influence. The organisation can only defend against hazards, it cannot remove them. Similarly, an organisation can only strive to minimise the incidence of unsafe acts, it cannot eliminate altogether the basic human propensities for committing errors and violations. All that an organisation can effectively do is to manage the internal factors known to play a part in creating unsafe acts.

Rather than striving to exercise direct control over the incidence of LTIs and accidents, organisational managers should seek to govern those internal factors relating to design, hardware, training, maintenance, procedures, goal conflicts and the like. These are the *manageable* factors determining an organisation's general "safety health", and it is these which should properly constitute the focus of safety management. But they belong to the hidden side of safety and can only be revealed by organisational indicators (i.e., Channels 4 and 5).

REFERENCES

Groeneweg, J. (1991) Controlling the Controllable. Ph.D. thesis. University of Leiden, The Netherlands.

Layfield, F. (1987) Report on the Sizewell B Public Inquiry. London: HMSO.

Mintzberg, H. (1979) The Structuring of Organisations. New Jersey: Prentice-Hall.

Reason, J. (1987) The Chernobyl errors. Bulletin of the British Psychological Society, 40: 201-206.

Reason, J. (1988) Errors and violations: The lessons of Chernobyl. In: E.Hagen (Ed.) 1988 IEEE Fourth Conference on Human Factors and Power Plants. New York: Institute of Electrical and Electronics Engineers.

Reason, J. (1990a) The contribution of latent human failures to the breakdown of complex systems. Philosophical Transactions of the Royal Society, London. Series B. 327: 475-484.

Reason, J. (1990b) Human Error. New York: Cambridge University Press.

Rolt, L.T.C. (1978) Red for Danger. London: Pan Books.

Wagenaar, W.A. and Groeneweg, J. (1987). Accidents at sea: Multiple causes and impossible consequences. International Journal of Man-Machine Studies, 27, 587-598.

Ward, D.A. (1988) Will the LOCA mindset be overcome? 1988 IEEE Fourth Conference on Human Factors and Power Plants. New York: Institute of Electrical and Electronics Engineers.

Westrum, R. (1988) Organisational and inter-organisational thought. World Bank Workshop on Safety Control and Risk Management. Washington, D.C., October 1988.

Woodham-Smith, C. (1953) The Reason Why. London: Constable.

Chapter 3

A FRAMEWORK FOR DESIGNING NEAR MISS MANAGEMENT SYSTEMS

Tjerk W. van der Schaaf
Department of Industrial Engineering and Management Sciences
Eindhoven University of Technology
Eindhoven, The Netherlands

In this chapter we will focus on the *contents and the design process* of systems aimed at the reporting, description, analysis and interpretation of near miss situations. Although certain aspects of the proposed framework for designing such a "Near Miss Management System" (NMMS for short) will be aimed at issues like implementation, maintenance and acceptability, the main overview of such organisational aspects will be presented by Lucas in chapter 11 of this volume. Chapters 3 and 11 may therefore be considered as "twin chapters". Both of these are critically reviewed in the concluding chapter 12 in the light of the case studies presented.

After listing a few *general functional specifications* for a NMMS design framework its *seven basic steps* or modules are introduced. Subsequently the implications of stressing the *different purposes* from Chapter 1 are discussed. An *extended version* of the framework is then presented related to its functioning at higher organisational levels than that of the basic version. The chapter is concluded by formulating a *response to Reason's (see chapter 2) critical remarks* of the usefulness of near miss reporting, and by summarising the different possible *uses of such a framework*.

1. GENERAL FUNCTIONAL SPECIFICATIONS

Four fundamental ideas or requirements are regarded here as functional specifications for designing a NMMS:
1. the *only* function of the NMMS should be to *learn* at an organisational level from the reported near misses;
2. its coverage of possible inputs and outputs should be *comprehensive*;
3. the "heart" of the NMMS should be a *suitable model of human behaviour* in a socio-technical system;
4. the NMMS should not be an "alien" system within an organisation, but be *integrated* where ever possible with other management tools.

1.1. Learning from near miss reports

Organisational learning should be central to the NMMS, that is: a progress-ively better insight into *system* functioning, not into individual performance. The final goal of the NMMS is *learning to control* or manage the safety aspects of system functioning irrespective of the specific individuals inter-acting with the system. Except for instances of sabotage, the NMMS output should never lead to "apportioning blame" to individual employees.

Another aspect of the NMMS as a learning instrument is the *self correcting* nature it should have: by building feedback loops into the NMMS it should be able to improve itself continuously.

1.2. Comprehensive coverage

The NMMS should be comprehensive in several aspects:
- it should be able to handle not only near misses, but also actual accidents, damages, etc., or be capable of being linked to an existing accident report-ing system;
- in its description and analysis it should pay attention not only to *negative* deviations from normal system performance like errors, failures and faults, but also to *recoveries*, the "positive deviations";
- it should focus not only on technical components and human behaviour as contributing factors to a near miss, but certainly also to organisational and managerial causes.

1.3. Model-based design

Following the previous point, ideally a complete socio-technical system model of the organisation involved should form the heart of the NMMS. Since such a model will not be readily available, at least a suitable model describing *indi-vidual* behaviour in a complex technical environment should be chosen as the "information processing part" of the NMMS. This model then dictates not only the required input data (taken from the near miss report) but also the methods of analysing and interpreting its results in terms of suggestions of specific measures to be taken by management.

1.4. Relationship with other organisational tools

The NMMS must be able to benefit from and contribute to other existing tools for measuring or understanding an organisation's performance, e.g. other safety-related information systems, audits, Total Quality programmes, etc. This also means that the amount of "success" of a NMMS should, in itself, be considered as an important measure of an organisation's performance or "safety culture" (see Chapter 11).

2.   BASIC NMMS FRAMEWORK

Figure 1 shows the proposed basic framework, consisting of seven modules which together should form the "building blocks" for different types of NMMS's.

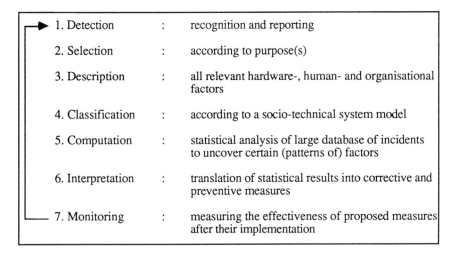

1. Detection	:	recognition and reporting
2. Selection	:	according to purpose(s)
3. Description	:	all relevant hardware-, human- and organisational factors
4. Classification	:	according to a socio-technical system model
5. Computation	:	statistical analysis of large database of incidents to uncover certain (patterns of) factors
6. Interpretation	:	translation of statistical results into corrective and preventive measures
7. Monitoring	:	measuring the effectiveness of proposed measures after their implementation

Figure 1. The seven modules of the basic NMMS design framework.

To explain the seven modules and their framework we will first describe the "information processing sequence" of near miss reports and subsequently the order in which these modules should be specified when designing a NMMS for a particular situation.

2.1. Processing sequence of near miss reports in the NMMS

2.1.1. The Detection module contains the *registration* mechanism, aiming at a complete, valid reporting of all near-miss situations detectable by employees.

2.1.2. A NMMS that works well will probably generate a lot of "deja vu" reactions on the part of the safety staff coping with a sizable pile of these reports. To maximise the learning effect some sort of selection procedure is necessary to *filter out the interesting reports for further analysis* in the subsequent modules. First of all, management objectives may of course lead to certain selection rules (e.g. special interest in personal injuries, or in product quality). Even more important however would be the presence of unique elements or unexpected combinations of elements, visible already by looking at the "raw" reports. Such reports would have to be ensured of the extra time and effort needed by the safety staff to apply all modules in these cases.

2.1.3. Any report selected for further processing must lead to a *detailed, complete, neutral description* of the sequences of events leading to the reported near-miss situation. For instance, an analysis based on Fault Tree techniques enables the investigator to describe all relevant system elements (technical failures, management decisions, operator errors, operator recoveries, etc.) in a tree-like structure. This tree will show all these elements in their *logical order* (by means of AND- and OR-gates) and in their *chronological sequence*.

2.1.4. Every element in such a tree may be classified according to the chosen socio-technical or human behaviour model, or at least every "root cause" (the end points of the tree) must be. In this way the fact that any incident usually has *multiple causes* is fully recognised: each near-miss report is analysed to *produce a set of classifications* of causal elements instead of the usual procedure of selecting only one of these elements as "the main cause".

2.1.5. Each near-miss tree as such generates a set of classifications of elements which have to be put into a *data-base* for further *statistical analysis*. This means that a NMMS is *not* meant to generate ad-hoc reactions by management after each and every serious near-miss report: on the contrary, a steady build-up of such a database until statistically *reliable patterns* of results emerge must be allowed in order to identify *structural factors* in the organisation and plant instead of unique, non-recurring aspects.

2.1.6. Having identified such structural factors (the *real* root causes), the model must allow interpretation of these, that is: it must *suggest ways of influencing these factors*, to eliminate or diminish error factors and to promote or introduce recovery opportunities in the human-machine systems and indeed in the organisation as a whole.

2.1.7. These suggestions to management will of course in practice be judged on other dimensions (e.g. time, cost) as well, but if they are accepted by management and actually implemented in the organisation they will have to be *monitored* for their predicted as opposed to their actual results, that is: for their *effectiveness in influencing the structural factors* they were aimed at. This may be done by the NMMS itself (see the feedback loop depicted in the 7-module framework): in the period following the introduction of the measures, near-miss reports should show a different frequency of occurrence for these factors. If a plant has one or more safety-performance measuring systems apart from the NMMS (like auditing-based systems) then some effect will probably be detectable by these independent indicators of safety also, depending on the degree of "overlap of content" between such separate systems and the NMMS.

2.2. Sequence of designing the NMMS modules

As mentioned already in section 1.3., the model of (human/organisational) behaviour which is chosen, becomes the heart of the NMMS: it directly defines the Classification-Computation-Interpretation group of modules which forms the "information processing section". This section in turn can handle certain types of input data, at specific levels of descriptive detail, and therefore defines the first three modules (Detection-Selection-Description). The information processing modules 4, 5 and 6 also determine the ways in which the accuracy of their predictions (that is: the effects of the proposed measures) may be monitored in practice (i.e. module 7).

The first *design* question therefore should be an organisational one: whether the model, at the heart of the future NMMS, "fits" the organisation's safety culture (see Chapter 11) or not. *Compatibility of model and culture* is absolutely essential for a successful implementation and acceptance of the NMMS later on.

3.   IMPLICATIONS OF DIFFERENT PURPOSES FOR NMMS DESIGN

In Chapter 1 three different purposes of near miss reporting have been mentioned: modelling (qualitative insight), monitoring (quantitative insight), and motivation (maintaining alertness). The reader will probably have noticed aspects of all three purposes in the description of the NMMS framework modules given in the previous section. In this section we will try to disentangle this inevitable but confusing listing of all seven modules in a "general" framework by indicating the subsets from these seven modules which are implied by focussing on *one* purpose at a time.

When the purpose is *modelling* we are only interested in reports of "new" near misses, selected from as complete a detection phase as possible, and subsequently described in great detail; classification and interpretation should be flexible enough to permit looking for "new" causal factors, which must be handled by new or existing measures.

Such new qualitative insights may then be formalised in the *monitoring* version of NMMS. This implies looking for known near misses only, which are routinely classified without any real selection or detailed description at all, and fed into an already existing database to detect any statistically significant trends in terms of improvements in error prevention or recovery promotion. The key issue here is to monitor whether the *existing* safety management measures are able to control the *known* hazards in the system.

The third purpose *motivation* is of a different nature than the preceding two because it describes not so much a safety management *tool* but rather a general condition or attitude which applies to *all* levels of personnel: the awareness that, in spite of all procedures, hardware precautions and training, the work environment still remains "dangerous" to a degree that it is sensible to follow existing safety rules even though "nothing" may have happened in a long time. The act of recognising and reporting near misses itself becomes an important *reminder* to act safely; also the selection of specific examples of old, well known but still recurring problems and new, "impossible" combinations of factors, described in detail in the setting of one's own workplace, should provide convincing illustrations of the fact that an absence of overt accidents is not to be equated with a perfect hazard control system. Instead, possible serious effects of seemingly minor deviations must be stressed, as should the factors and measures which have acted as effective recoveries.

For the sake of clarity the points raised above are summarised in figure 2, indicating which modules are or are not implied (in a very global way) by the three purposes mentioned.

Purpose of near miss reporting

	Modelling	Monitoring	Motivation
1. Detection	everything	known problems only	recognising and reporting
2. Selection	new reports only	absent	convincing, detailed examples of new and old hazards
3. Description	detailed	absent or very superficial	
4. Classification	flexible: looking for new root causes	routine: standard set of root causes	absent
5. Computation	absent only single events considered	periodic analysis of updated large data-base	absent
6. Interpretation	finding (new) ways of improving prevention and recovery	absent already prescribed by module 4	near misses as precursors; focus on recovery mechanisms
7. Monitoring	absent	comparing pre-dicted and actual effects of imple-mented measures	absent

Figure 2: An overview of different versions of the basic framework of fig. 1, according to different purposes.

In practice however the usual case will be that of a *combination* of two or three purposes into a single system. This makes the distinctions in figure 2 more illustrative of the use(s) of the modules themselves than as guidance for a realistic NMMS design task.

4.   EXTENDED NMMS FRAMEWORK

The basic version of the NMMS framework described in sections 2 and 3 shows it functioning at the *level of the safety department* in an organisation. The learning process thus takes place at the level of "end-users" (e.g. operators, train drivers, etc.), their direct supervisors and the local safety staff. *Feedback loops* which make this learning process possible are not only the "monitoring" loop from module 7 (or 6) back to module 1, but also several smaller loops within the framework: e.g. in modelling module 6 may very well influence module 4, which in turn may change the ways in which the "input" modules 1, 2 and 3 operate.

At higher organisational levels however important extra feedback loops are necessary, leading to an *extended version*: Detection of "impossible" events or classification of "new" root causes may lead to direct inputs to the *engineering department* for hardware solutions (including ergonomic improvements of the human-machine interface). *Operations management* may also have to react to such inputs by changing the worksituation (e.g. staff levels, task allocation, communication channels, etc.). Finally, at the *senior management level*, sometimes far-reaching re-evaluations of the balance between production, safety and environmental priorities will have to be made. Also major changes relating to NMMS's own performance and its mixture of purposes will by definition mean that "outside" loops will be needed for such decisions (see also Chapter 12 for a discussion of the NMMS life cycle).

5.   USEFULNESS AND USES OF THE NMMS FRAMEWORK

In this section we will first respond to Reason's critical remarks made in the previous chapter and then conclude the chapter by summarising the possible uses of the NMMS framework.

5.1. Response to Reason

In Chapter 2 Reason has launched several fundamental critical remarks concerning the usefulness of current near miss reporting schemes. We fully agree that incident and accident reporting systems are by themselves *insufficient* to support effective safety management. Indeed, this is true of any safety management tool or approach, considering the enormous complexity and variety of accident-causation processes. Even proactive methods based on well-documented research like Reason's General Failure Types auditing tool will not succeed in providing all the answers to the questions triggered by the different safety management purposes of Chapter 1.

It is also as yet not proven that *backtracking* from incidents or near misses necessarily stops at the level of "tokens": the extended version of the *model-based* framework in section 4 clearly *aims at "function types" and even "source types"*. Nor is it a real problem that specific connections for a single near miss cannot reliably be traced back to "latent pathogens", because the essential feature of the monitoring version of NMMS is precisely the *analysis of large data-bases* of incidents and near misses in order to identify *statistical* patterns of error and recovery factors (module 5).

Finally, it is concluded in Chapter 1 that near miss reporting may serve three different, valuable safety management *purposes*: monitoring, modelling, and motivation. The same cannot be said of *auditing tools* (like Reason's) which seem to aim largely at *monitoring*, and possibly *motivation*, without (any major) benefits for modelling purposes. Such sophisticated auditing tools may indeed provide reliable and valid indices of an organisation's intrinsic safety state, but they are not comprehensive in the sense that they fulfil all three purposes to a sufficient degree.

5.2. Possible uses of the framework

Summarising the main points of this chapter we can distinguish the following ways in which the NMMS framework may be used:
- the simplest form is to use it as a *checklist for describing* the status of accident/ incident/near miss reporting systems. In this way a complete inventory of such a system is made by simply "following" a near miss report being handled by the existing information processing sequence in a chronological order. Examples in this volume are given in chapters 6, 7 and 8;
- secondly we may regard the framework as a *normative model* for (re) designing such systems. Having described an existing system in the way mentioned above, immediately "missing" modules and reversals in the sequence of modules may be noted by comparing the described system with the normative framework. Of course it may also serve as a guideline for designing a completely new reporting system, in the order described in section 2.2. (see Chapter 6 for an example);
- finally, by taking its use as a descriptive checklist as a starting point, it may become a framework for *designing NMMS support systems*: system documentation, training programmes and decision support for learning how to use it, and the explicit design of the feedback loops in- and outside the NMMS itself (see section 4) may be guided by it.

Chapter 4

UNDERSTANDING, REPORTING AND PREVENTING HUMAN FIXATION ERRORS

M. Masson
Commission of the European Communities
Joint Research Centre
Ispra (VA), Italy[1]

1. INTRODUCTION

The behaviour of the human operator remains a central component of the safety of any complex socio-technical system, such as electrical energy production, chemical or petrochemical industries and transportation systems.

Accidents fortunately occur with less frequency than both unsafe situations and human errors. In addition, we can learn as much from the recovery from an error as from its consequences. Hence there is a need to extend the scope of safety studies to everyday work situations.

Much critical real-world problem solving behaviour takes place in dynamic, event-driven environments, where partially ambiguous or conflicting cues appear gradually and where situations evolve in both an intrinsic way and in relation to the actions taken by operators. These situations, in which people have to amass, integrate and make decisions based on uncertain, incomplete and changing evidence, arise easily at all organisational levels.

In these complex environments, case studies have shown that a major error pattern is *fixation*, a failure to revise situation assessment with new evidence (De Keyser and Woods, 1989). Such errors are initially quite difficult to identify. They become manifest later in the evolution of the event, when people do not succeed in updating their view of the world, in spite of cues that can in retrospect be recognised as counter-indications.

Can we understand the nature of these perseverance errors or are they only a matter of chance? And, if we succeed in identifying their essential character, what kinds of practical applications would such knowledge bring? Might a better understanding of human error help to specify new data reporting systems or to improve existing ones? Can we foresee improvements in domains such as error prevention and training?

In this paper, COSIMO, the cognitive operator model currently implemented at the Ispra JRC, is described, with particular emphasis on its theoretical foundation. Then we show how this modelling approach can be applied to analyse one industrial case study of a human fixation error.

---

[1] The author has now moved to: Université de Liège, FAPSE, B32 Sart-Tilman, B-4000 Liège 1, Belgium.

The aim is to identify those conditions under which our simulated operator starts to fixate and the ways of producing error recovery.

The COSIMO modelling programme currently allows us to reproduce accident cases or to simulate new ones by generating complementary data on the aetiology of accidents. But extensions can be foreseen. The following section presents the relationship between this ongoing project and the refinement of accident and near miss reporting systems.

The methodology applied is cyclical. Reported accident cases and cognitive theory give the model its psychological and practical background. Conversely, the analysis of virtual incident cases produced by simulation leads to the identification of factors that explain such simulation profiles.

The main assumption is that those identified data types could also typically be used to refine reporting systems, especially concerning the decisions and actions taken by the actors just before the incident. Improved reporting systems would then provide researchers with improved data, which in turn would help to improve simulation. And so the cycle starts again. The expected gain, in terms of information quantity and quality, is far greater when reporting is extended from accidents to near misses.

The later sections of the article provide some recommendations in two major domains: the prevention of human errors and the improvement of operator training. Guidelines are given for safety specialists to minimize the influence of identified error releasing factors, taking into account the way operators are expected to behave under related conditions.

Finally, the last section presents some conclusive remarks and outlines the direction of future research.

2.   DEFINITION OF FIXATION ERRORS

As defined by De Keyser and Woods, fixation error is "a failure to revise a situation assessment as new evidence comes in". This cognitive encystment - as it could be called - has been widely recognised as a major source of human error in dynamic and complex environments (De Keyser et al., 1981; Woods, 1984; Woods et al., 1986; Cook, Mc-Donald and Smalhout, 1989).

In a fixation, the initial problem solving context has captured and biased the problem solver in certain directions. It is well known by psychologists that the initial characteristics of an accident tend to activate related knowledge structures, which in turn affect the way people select, perceive and interpret data and understand the ongoing situation (Kahneman and Miller, 1986). This has been reported especially during the early stages of the accident sequence : the operator becomes locked into one view of the world, one strategy and one goal, and completely refuses to change. One can observe great persistence in his behaviour, repeating or carrying on the same action(s) without making any progress (Daniellou, 1982; Fichet-Clairefontaine, 1985). This behaviour is easy to detect, because of an impressive sequence of quasi-compulsive repetitions, despite an absence of results. The operator is often aware that the system does not react according to his expectations, but does not succeed in updating his strategy. Moreover, the erroneous actions he takes could shift the system to a degraded state, which is far more difficult to identify and manage due to second order complexity effects.

It is mainly to escape from this cognitive bias that it has been proposed to resolve a fixation case by adding a "neutral observer" to the team. This neutral observer, an expert operator in the field, should not be present at the beginning of the accident. He will then have the advantage of not having been involved with the formulation of any initial anchor hypothesis when tackling the problem.

3.   A CASE OF FIXATION ERROR COMING FROM INDUSTRY

This brief sample fixation case comes from an industrial application. It has been reported by De Keyser some years ago (De Keyser et al.,1981; De Keyser and Woods, 1989).

In this case, an operator was monitoring the treatment of recoverable materials during a night maintenance shift. The operator's task was to send materials so as to fill a set of tanks in succession: once a tank is filled the following one would be connected and so on. The switch-over operation was ordered manually by the operator, without the assistance of automatic devices. It was an unusual procedure for effluent treatment, which this operator had only carried out twice, both times under standard conditions. Usually, the flow rate for this filling process is 35 cubic meters per hour, and the operator had a strong expectation about the time required to fill a tank at this rate. What the operator did not know was that the flow had been increased to 55 m^3/h by an engineer just before he came on duty.

This change had been reported in a book, but the operator did not notice the message nor was he explicitly informed about it. As the crew is reduced during the night, the operator was alone when he entered the control room around ten p.m. Shortly after midnight, warning signals on level and pressure began to indicate tank overfilling. The operator noticed them but did not respond in the appropriate way. Instead, he carried on his ongoing procedure assuming the normal conditions were valid. So he waited until the usual time had elapsed before shifting the flow to another tank. This procedure was repeated for a second and third tank, with a global duration of about six hours!

The post-incident inquiry suggested that the operator behaved this way for several reasons. He knew that the alarms he receives on level were only "very high" alarms - as opposed to "overfilling" alarms - meaning that a tank is 80% full, and thus did not require any particularly fast response. The alarms occurred before the expected time to fill one tank had elapsed. So he believed sufficient volume was still available and postponed flow shifting. His conviction about the state and evolution of the process was stronger than external disconfirming cues. To make the situation cognitively coherent, the operator assumed that the alarms did not effectively reveal the behaviour of the system but were due to instrumentation failures. This attribution persisted until alarms triggered by overfilling disturbances alerted other operators outside the control room.

4.   COSIMO: A COGNITIVE SIMULATION MODEL

4.1. Computer implementation and technical background

How can we analyse this report of a fixation error using cognitive psychology and cognitive modelling?

The approach to the modelling of man-machine systems that we are following at the Joint Research Centre of Ispra is a simulation methodology, where human behaviour and technical system responses are modelled at a theoretical level and integrated for studying interactions during transients and accidental conditions. The operator model is called COgnitive SImulation MOdel (Cacciabue, Decortis, Mancini, Masson and Nordvik, 1989), a software system dedicated to simulating the operator's cognitive processes and thus able to produce "human-like" errors.

It is written in common-lisp on a 3640 SYMBOLICS computer. This AI language is particularly adapted for simulating human problem solving and decision making. Moreover, the use of a recently implemented software package, KEE ("Knowledge Engineering Environment") (Fikes and Kelher, 1985), allows easy representation of operators' knowledge and know-how. This shell also provides good graphics facilities to interface with the model.

The human model interacts with models of several subsystems of a typical Nuclear Power Plant, e.g. the Auxiliary Feed-Water System, implemented on a network of SUN computers. Currently the model is implemented for the application area of nuclear energy production, however, other industrial situations could also be modelled, providing that they share similar cognitive activities (Cacciabue, Decortis and Masson, 1988).

COSIMO is currently operational. One basic application is to explore possible patterns of human behaviour in simulated accident situations and thus to identify suitable safety counter-measures. COSIMO can be used to:

- analyse how operators are likely to act given a particular context,
- identify difficult problem solving situations, given problem solving resources and constraints (operator knowledge, man-machine interfaces, procedures),
- identify situations that can lead to human error and evaluate their consequences,
- identify and test conditions for error recovery,
- investigate the effects of changes in the man-machine system.

In the context of this particular study, the aims are to reproduce, analyse, prevent and mitigate human fixation errors.

4.2. Immediate versus deep reasoning

The following discussion - a rather theoretical one - is presented in order to stress what sorts of human features the COSIMO simulation programme has to model in order to ensure its "human character" and to be able to produce realistic human errors.

In emergency or critical situations the operator may be stressed, anxious or tired and has to find an appropriate solution in a very limited time. According to our cognitive approach, in such a situation, no sophisticated reasoning activity is performed and the operator resorts to previous experience of similar cases. We identify this mental process as a kind of immediate reasoning, which uses immediately accessible chunks of knowledge, called rule-based frames, having the form: "When this situation occurs, just apply this recovery procedure". Their accessibility in mind is positively related to their frequency of use and to their success in the past. Observations show another positive relation between frequency and knowledge extent: the higher the subjective frequency of a certain event, the more documented and relevant will be the corresponding knowledge available for recognition.

Behaviour is based on pattern recognition: to assess the state and short term evolution of the process, the operator simply selects and picks up significant details in his environment and matches them against expectations provided by his work experience. As soon as the situation is recognized, or believed to be so, the actions to be taken are provided by the attached written or mental procedure. No strategy nor action plan has to be worked out on the spot. The operator simply knows what to do. He behaves in a kind of auto-pilot mode.

But not all problem solving activities involve short-cut reasoning. Sometimes, especially when he is novice in a field, the problem solver does not have prior experience to rely on. When no appropriate explanatory hypothesis or recovery strategy can be found, the operator has to elaborate a strategy on the spot, using a mostly serial process of inference called deep reasoning, which requires more effort. Problem solving processes rely on the combination and exploitation of knowledge-based frames, that contain physical and engineering beliefs constituting the operator's mental models (Moray, 1987). These representations concern the way he understands the structure and behaviour of the physical system, according to the tasks he has to carry out. These beliefs can be described as: a set of knowledge about the system's lay-out, structural and functional properties; physical laws expressed as operative relations between physical variables and system behaviour, and temporal dependencies between process states, control variables, action ordering and scheduling. They are mostly causal in nature and are also indexed according to their frequency of use and success in the past (Masson and Cacciabue, 1989). Reasoning at this level is slow, tedious and effortful. It demands full concentration.

4.3. Models of cognitive activities

Psychological analyses of operators at work lead us to subdivide their behaviour into cognitive activities, like diagnosing, monitoring, scheduling, etc. These activities are related and thus partially interdependent. For example, an operator's perception is calibrated by his present view of the world, the way he understands problem evolution. But in turn, being selective, perception tends to shape and sustain the current hypothesis against disconfirming cues. Perception and performance determine the content of knowledge acquisition and the way knowledge is organised by the operator. This knowledge in turn largely determines data selection and interpretation and, of course, diagnostic efficiency.

In the present version of COSIMO, five interrelated cognitive activities have been modelled and implemented.

Data Decoding

The communication between the model of the physical system and the model of the human operator is established by sending files of coded numerical data concerning all indicators and alarms present in the machine model. COSIMO has its own internal representation of the control panel. The first processing step aims at decoding this received data and updating the corresponding indicators in COSIMO's control panel representation. Strictly speaking, this first step is not cognitive.

Perception

Not all the data present on the control panel are simultaneously perceived by an operator. This should be reflected in the simulation. Data selection is performed by means of a cognitive filter. Its function is to decide, among all the data received and decoded, which items will currently be taken into account or perceived by the model. Two basic quantitatives are computed: physical salience, that reflects in a synthetic way the perceptual properties of the cues (e.g. type of indicators, position in the control panel, surface, light or sound intensity, etc.) and cognitive salience, expressing the subjective familiarity of the cues, which is positively related to their frequency of encounter in the past.

The cues are selected on the basis of a combination of the preceding scores: all cues whose salience reach a given filter threshold are selected. This threshold has a system default value, but this value may be modified during a simulation.

Data Interpretation

Only those data that are selected by the cognitive filter are interpreted. Interpretation is a cognitive process by which numerical data are translated into semantic qualitative or ordinal values, using reference distributions. The interpreter has a variable sensitivity, in order to duplicate the effect of various cognitive states like tiredness or attention loss.

Similarity Matching

Once the cues have been decoded, selected and interpreted, they are available to the working memory for matching. The number, form and quality of these cues greatly determines the response of the model's knowledge base: the more pertinent the calling conditions, the less numerous and the better supported will be the candidate explanations. Not all the knowledge base is accessed. Once a hypothesis is instantiated in the working memory, it will in turn govern all the behaviour of the model: all cues are priority-matched against this hypothesis. Secondly, search is restricted on the knowledge base's temporal axis. As soon as the model is synchronized to the event evolution, only that part of the knowledge corresponding to the relevant time span is available for matching. These options are default ones. All kinds of modifications can be introduced from both inside and outside the simulation.

Frequency Gambling and Selecting

In the case of ambiguity, either in the interpreted data or in the symptoms present in the knowledge base, more than one explanatory hypothesis may be "brought to mind" by Similarity Matching. The Frequency Gambling primitive has to solve competition or conflict between partially matched hypotheses. Firstly, matches of value and frequency are combined to deliver one final support to each hypothesis. This combination may take various forms depending on the choice of the user or on the current cognitive state of the model. The hypothesis selected is the one obtaining the greatest support.

These two primitives act so as to minimise cognitive strain and to maximise the chances of automatic pattern recognition, which is considered the most common strategy - and thus the default one - by which people tackle problem solving (Reason, 1986, 1987).

Hypothesis Evaluation

The hypothesis selected is the hypothesis with the highest final support. This support can however be too low to give the model any confidence in this candidate explanation. An evaluation stage aims at deciding if a potential candidate explanation can be trusted or if it must be rejected, allowing another knowledge base search to be initiated.

Execution

Once an explanatory scenario has been selected and the situation has been recognized and "understood" by the model, the corresponding recovery procedure has to be implemented. The execution module has to decide when to check the system's evolution and what actions have to be taken to monitor the situation.

Control Architecture

The selection, sequencing and coordination of these various models is managed by an independent programme.

Various parameters are in charge of constraining and tuning the behaviour of the main lisp functions, allowing the system to reproduce subtle variations in the overall model's performance (Woods, Roth and Pople, 1987), so as to explore the influence of various cognitive states, like stress, tiredness or excessive caution, on the evolution of the simulated incidental sequences.

5.   A FIXATION CASE AS REPRODUCED BY COSIMO

As a particular application, a generic case of fixation error has been reproduced by simulation.

The case run was different from the study case presented in section 3, at least as far as the technical system involved is concerned. In this simulation, a case of nuclear station blackout was simulated. Under these conditions, the operator must switch on an auxiliary heat removing system called the Auxiliary Feed-Water System (AFWS). One should keep in mind that this event is also quite unusual.

COSIMO has a set of memorized descriptions summarising the system's physical evolution and the appropriate monitoring procedure under various normal and incident conditions - AFWS operation is per se an abnormal event but secondary incidents like a pump or a valve failure may further degrade the situation.

According to De Keyser (private discussion), the accumulation of these two unusual (and unexpected) events can be seen as the main cause of fixation errors; she proposes the term *change in change* to describe the combination. Any unnoticed change in change may overload the capacity of an operator to manage an incident. Such a condition, in any case, greatly increases the chance of erroneous behaviour.

We simulated a case where, in the beginning, the AFWS was functioning perfectly and COSIMO did succeed in identifying the correct corresponding scenario through the Similarity Matching and Frequency Gambling mechanisms. This success in coping with an unusual event then greatly reduced its field of attention and its caution in evaluating its own behaviour. Having got the right explanation, it then only had to implement the previous actions step by step, at the right moment.

The main error-inducing factor here is subjective frequency. As the simulated event is quite rare, once it does appear the likelihood of alternatives (like degraded versions, which are far less frequent still) is very low. This problem was still more evident in the tanks incident, where among the set of similar tasks that the operator had already encountered, both of them were exactly the same. The operator never had the opportunity to infer any variability across cases.

But in our simulation, after 300 seconds, the simulation driver inserted a failure into the system, breaking the inlet of one of the four Steam

Generators. COSIMO did not notice the change. Giving low attention to the indications it was receiving on the AFWS abnormal behaviour and being very satisfied about its current hypothesis, it carried out the initially selected strategy without any attempt to update it in the light of disconfirming cues: COSIMO was in fixation!

Technically, this behaviour was produced by using a special combination of control parameters. The effects of this selection on the performance of the model can be summarized as follows:
- by-passing any deep reasoning attempt;
- narrowing the model's field of attention by increasing its filter threshold;
- increasing the weight of frequency relative to degree of matching in the computation of a hypothesis' final support;
- decreasing the evaluation threshold, thus making the model less cautious about mismatches between hypothesis-related expectations and interpreted cues.

Such settings carry more than just theoretical interest in simulation research. The know-how they reflect in terms of cognitive modelling leads to insights about how to improve the analytical power of incident reporting systems.

What do we learn from simulation? How does it help in analysing reported cases coming from real work environments? It is interesting to go back to the case study presented in section 3 and look for similarities.

In section 3, it was stated that the procedure the operator had to carry out for effluent treatment was an unusual procedure. In fact, the operator had only followed it twice in the past. A second important point is the fact that an additional change had been introduced in the environment, without explicitly informing the operator.

Once again, the notion of change in change could also be proposed in this case. The report also highlights the influence of the procedure content and, particularly, of its informal version: the operator based his behaviour on the time that he believed to be necessary to fill the tanks. This mode of monitoring the task was obviously extraneous to the formal, written procedure and can be seen as a sort of hidden agenda.

The last main aspect revealed by this case study is the importance of the operator's current view of the world and of its associated recovery procedure: once the operator had the conviction that he was behaving properly, he became resistant to any attempt at change, exactly like it was obtained in the simulation.

The important point to highlight here is that the way of understanding and analysing this incident report is directly provided by the theoretical background and the technical settings used in simulation.

6.   COGNITIVE MODELLING AND IMPROVEMENT OF REPORTING SYSTEMS

The methodology followed in this research programme is an interactive one. From the analysis of accidents and, more specifically, by focussing upon the behaviour of front line operators at the extreme end of the accident causal chain, we get unique information concerning the main features and biases characterizing human behaviour in real work situations, as well as the role played by the context, interface and environment.

Conversely, once those data have been programmed in a computer model, variations and new cases can be produced in a fully symbolic way and analysed deeply, without having to wait for the occurrence of such variations.

The analysis of simulated human error cases contributes to improving our understanding of the conditions triggering human error, both in the operator and in the environment.

Insights gained through such analyses help to identify the cues which can be the origins of system's drifting from a nominal to an accident state. Thus, we can identify the kinds of data we need to collect to refine reporting systems especially focussing on the decisions and actions of operators.

Improvements provided by such modelling effort are firstly dedicated at reporting and preventing human active failures, i.e. errors and/or violations committed mostly by operators maintaining or controlling the system, and whose effects are felt immediately. This methodology will not produce the same kind of insights in the area of the organisational, sociological and communication factors committing to accidents, whose importance has been highlighted in other papers (see for example Reason, 1989, and Chapter 2 of this volume).

Improvements are not expected to be limited at accident reporting systems: when properly designed, any system of that kind could be extended to the registration of incidents and near misses. The amount of information provided by near miss reporting would, in turn, greatly contribute to enlarging and refining our knowledge of the way humans behave and how they succeed in avoiding accidents in complex systems. Those refinements would then help to improve modelling in a kind of self-improving closed feed-back loop. But what we gain from simulation runs and case studies is not ultimately aimed at improving simulation as such.

At every stage of development, the knowledge of the various factors that shape COSIMO behaviour, like time pressure, the extent and quality of its knowledge and of the cues it receives, can also be used to develop safety recommendations and to improve training programmes.

7.   MITIGATION OF FIXATIONS AND PREVENTION OF HUMAN ERRORS

As safety specialists, we have to think about the main characteristics of human information processing in order to improve operators' efficiency, to minimise mental workload and to prevent human errors.

Case studies - such as the one presented in this paper - help to capture the basics of operator behaviour under abnormal conditions. The question is: given that we have a reliable representation of the way humans operate, how can we maximise the productivity and reliability of this mental processing?

In this section, we do not intend to be exhaustive. The purpose is simply to identify suitable safety recommendations in the light of the model. As a model is always reductive, it will always restrict the field of potential investigations. However, it has the advantage of proposing a means of deepening and testing some hypotheses in a symbolic way: once a hypothesis is implemented, all modifications can be run, fully controlled and analysed very rapidly.

As mentioned before, this modelling project has mainly focussed on immediate reasoning, which is the default way people cope with problems. Some basic components of the model will be reviewed with the objective of increasing overall efficiency. A more general and complete set of safety measures regarding fixation errors may be found, for example, in De Keyser and Woods (1989).

7.1. Knowledge enlargement

Operators have to recognize and recover from faults and failures occurring in the technical system they manage. To be successful in this task, they should have a reliable and extensive knowledge base of prototype cases to refer to. The first recommendation is thus to enlarge operators' know-how by increasing the number of such ready-to-access reference cases (rule-based frames) and by completing and refining their understanding of the physical system (knowledge-based frames). Such objectives are usually taken into account in classical training.

7.2. Knowledge and interface improvement

What the model shows is that when the knowledge base grows, the number of potential candidates likely to be brought into the working memory as the product of Similarity Matching, i.e. the number of hypotheses or scenarios between which the operator would hesitate, greatly depends on the quality of the available knowledge and the quality of the cues selected in the environment. Each description and each item of environmental data should thus be as unambiguous as possible. This can be achieved by increasing the discriminatory power and salience of the knowledge content and the control room indications.

In other words, to increase efficiency, the knowledge an operator has about the system, the procedures he has to carry out and the cues provided by the interface have to be as differentiated as possible, so as to avoid confusions.

This is a way of reformulating an old but unfortunately too rarely applied design golden rule: spot conflicts between similar views of the world and action routines and make them as different as possible.

8. COGNITIVE MODELLING AND TRAINING IMPROVE-MENT

8.1. Frequency and training

As described in section 4.2., immediate reasoning is the default way by which operators tackle problem solving. Operators first try to acknowledge the situations they are confronted with by comparing them to prototypical cases they have in mind. The subjective frequency of such previously experienced reference cases thus also has a crucial impact on the efficiency of that recognition process.

Training may be seen as a concentration of a lifetime's experience: everything interesting may be experienced in a short period of time. Three guidelines for selecting training scenarios are proposed:

(1) Training should somehow reproduce the objective frequency of occurrence of the study cases in real environments. This will give coherence to the operator's knowledge. To become operational rapidly, the most common cases are studied first and are best documented.
Trainers should also be aware of sampling problems. For example, if five exotic birds are intensively studied during a training session and if four of them happen to be of the same colour, one could hastily extrapolate the colour of all exotic species from this limited sample. People are very prone to generalize from non-representative samples (Kahneman, Slovic, Tversky, 1982).

(2) Not all accidents have the same consequences. Some produce large scale catastrophes, others are benign. One should therefore spend time and energy in making people able to manage the most dangerous scenarios even if they are not among the most frequent ones. When one uses the word "frequency" in this context, the key is to acknowledge that this factor does not exactly duplicate the objective frequency of the cases concerned. A better formula would be *frequency-in-mind*, which is also particularly related to factors such as the gravity of plausible outcomes, the time spent in learning or managing the cases, and the emotional charge involved.

(3) Special attention should also be given to rare events. Frequent cases are likely to be met during everyday work. Repetition will consolidate their cognitive availability. Rare events, by contrast, are in general never encountered after training. But what can occur if they actually appear? We have to remember that the lower the frequency-in-mind, the poorer the knowledge available to overcome the problem. Moreover, it is well known that unpracticed knowledge is bound to vanish. Here we see the need for access to complementary documentation, for example by interacting with intelligent decision support systems (this argument is beyond the scope of this paper).

8.2. Training in dynamic worlds

Training sessions should duplicate problems as closely related as possible to those occurring in the control room. One dimension should imperatively be maintained: time (Decortis and De Keyser, 1988). Accidents are not local events. In a slowly responding system, they sometimes evolve over hours before being recovered. In dynamic worlds, there is no "one shot" well-formulated diagnosis of the situation. Rather, people have to update their initial situation assessment, as mismatches are detected between expectations and new evidence. Furthermore, diagnosis, plan updating and planning are not separate sequential stages but are deeply interrelated processes, where feed-back plays a central role (Woods and Roth, 1986). Operators should therefore be prepared to tackle that temporal dimension, by managing dynamic cases during training sessions. Thus they will have the opportunity to develop suitable dynamic chunks of knowledge, concerning the system's typical evolutions, under both normal and degraded conditions.

8.3. Making operators aware of their strengths and weaknesses

Human immediate reasoning is rather efficient. It usually performs very well in standard, common situations and keeps mental workload manageable. It is also unfortunately error-prone.

During problem solving, once a candidate (whether an explanatory hypothesis or a recovery strategy) is believed to be the right one, once the situation has been "framed", the operator often increases his confidence in that candidate. All the cues coming from the environment or even generated by reasoning are accounted for and interpreted in the light of this currently active view of the world. Attention is paid to supporting data, and contradictory cues are discarded. At a later stage, contradictory evidence may be completely denied! This seriously impedes any attempt at recovery. This ongoing process, called *confirmation bias*, finally leads to the operator's "cognitive lockup". Operator training should therefore highlight this powerful, recurrent bias. The objective is to prevent human fixation errors. Operators should be trained to break, early on, this unsatisfactory immediate reasoning loop before being locked in it and losing any "cold reasoning" ability.

9.   CONCLUSION

The development of new AI techniques and the recognition of the role played by psychological aspects in the management of complex plants, have shown the need to implement sophisticated models able to duplicate the way operators behave and reason, in relation to the physical systems they are in charge of (Rasmussen, Duncan and Leplat, 1987). In particular, human factors considerations and evaluations of engineered systems by means of Man-Machine Interaction methodologies have become key issues, and progress in this field could greatly contribute not only to a deeper understanding of operator behaviour, but also to the development of Intelligent Decision Support Systems (Hollnagel, Mancini, Woods, 1986).

In this context, Man-Machine Interaction is generally recognised as a central issue in the design of a complex industrial system, and different research areas have been involved in this topic, namely: Artificial Intelligence techniques for the design of decision support systems, ergonomics for the design of control rooms, and reliability and human factors for safety studies focussing on the new role of operators and on human errors.

In this paper, the recent development of a cognitive model, based on the interaction of engineering methodology, psychological approaches and Artificial Intelligence tools, has been presented in some detail.

The practical interest of the COSIMO research programme is not bounded at improving data collection. In this article, elements of analysis and recommendations have been formulated in two important domains: the prevention of human errors and the training of operators, in the light of what has been learned from simulations. Emphasis has been given to concepts like quality and discrimination power in the contents of both operators' knowledge and information system. The influence of subjective frequency in the operator's behaviour has been reported. Confirmation bias and its pernicious influence has been reviewed.

COSIMO is an ongoing project and applications are still in an early stage. In particular, the work to be done in cognitive modelling remains considerable in all fields of research concerned, from psychology to knowledge engineering and computer science. As a consequence, only the basics of operator behaviour are currently accounted for. However, it is not necessary to wait for the final stage of development before considering practical applications.

As a guarantee for the future, the mutual understanding, quite vivid nowadays, and the commonality of language and tools between engineers and psychologists are a promising basis for successful applied research into systems safety.

ACKNOWLEDGEMENT

The COSIMO project is a team enterprise. The author would like to thank James Reason, Pietro Carlo Cacciabue, Françoise Decortis, Bartholome Drodzdowicz, Giuseppe Mancini and Jean-Pierre Nordvik for their essential contribution to this project development.

REFERENCES

Cacciabue, P.C., Decortis, F. and Masson, M. (1988). Cognitive models and complex physical systems : a distributed implementation. 7th European Conference on Human Decision Making and Manual Control, Paris, 18-20 October 1988.

Cacciabue, P.C., Decortis, F., Mancini, G., Masson, M. and Nordvik, J.P. (1989). A cognitive model in a blackboard architecture: synergism of AI and psychology, Proceedings of the Second European Meeting on Cognitive Science Approaches to Process Control, Siena, Italy, October 24-27, 1989.

Cook, R.I., Mc-Donald, J. and Smalhout, B. (1989). Human error in the operating room : identifying cognitive lockup. Manuscript submitted for publication.

Daniellou, F. (1982). Stratégie de résolution d'incidents aux presses auto-matiques: le poids de la technologie et de l'organisation du travail. Communication au XVIIIème Congrès de la Société d'Ergonomie de Langue Française (SELF), Paris.

Decortis, F. and Keyser, de, V. (1988). Time : the Cinderella of Man-Machine Interaction. Proceedings of the 3rd. IFAC/IFIP/IEA/IFORS Conference on Man-Machine Systems, June 14-16 1988, Oulu, Finland.

Fichet-Clairefontaine, P. (1985). Etude ergonomique de l'influence de la conception de la salle de commande, de la stabilité du processus et de la diversification de la production sur l'activité des opérateurs dans quatre unités à processus continu. Thèse de doctorat en ergonomie, non publiée.

Fikes, R. and Kelher, T. (1985) The role of frame-based representation in reasoning. Communications of the ACM, September 1985, vol. 28, No 9.

Hollnagel, E., Mancini, G. and Woods, D.D., (Eds) (1986). Intelligent Decision Support in Process Environments. NATO ASI Series, Springer-Verlag, Berlin, Germany.

Kahneman, D., Slovic, P. and Tversky, A. (Eds.) (1982). Judgement under Uncertainty: Heuristics and biases. University of Cambridge, NY.

Kahneman, D. and Miller, D.T. (1986). Norm Theory: comparing reality to its alternatives, Psychological Review, 93, 136-153.

Keyser, de, V. et al. (1981). La fiabilité humaine dans les processus continus, les centrales thermoélectriques et nucléaires. Rapport CCE DGXII, CEERI, Bruxelles.

Keyser, de, V. and Woods, D.D., (1989). Fixation Errors in Dynamic and Complex Systems. In : Advanced Systems Reliability Modelling, Colombo, A.G. and Micenta, R., (Eds.)., Kluwer Academic Publishers, Dordrecht (forthcoming).

Masson, M. and Cacciabue, P.C. (1989). The contribution of cognitive modelling to the development of training technology and safety improvement. International Meeting on Education and Training for Prevention, Paris, 31 May-2 June, 1989.

Moray, N. (1987). Intelligent aids, mental models, and the theory of machines. international Journal of Man-Machine Studies. Vol. 27, 5-6, 619-630. London, Academic Press.

Rasmussen, J., Duncan, K. and Leplat, J., (Eds.) (1987). New Technology and Human Error. J. Wiley and Sons, London, U.K.

Reason, J.T. (1986). Recurrent errors in process environments : Some implications for the design of Intelligent Decision Support Systems. In: Intelligent Decision Support in Process Environments, E. Hollnagel, G. Mancini and D.D. Woods (Eds.), NATO ASI Series, Springer-Verlag, Berlin, Germany.

Reason, J.T. (1987). Modelling the Basic Error Tendencies of Human Operators, 9th SMIRT Post-Conference Seminar on Accident Sequence Modelling: Human Actions, System Response, Intelligent Decision Support. In : G. Apostolakis, G. Mancini and P. Kafka (Eds) Accident Sequence Modelling: Human Actions, System Response, Intelligent Decision Support. North-Holland, Amsterdam, the Netherlands.

Reason, J.T. (1989) The contribution of latent human failures to the break-down of complex systems. Presented at the Royal Society Discussion Meeting on Human Factors in High-Risk Situations, London, 28-29 June, 1989.

Woods, D.D. (1984). Some results on operator performance in emergency events. Ergonomic Problems in Process operations, D. Whitfield (ed), Inst. Chem. Eng. Symp. Ser. 90.

Woods, D.D., Roth, E.M. and Embrey, D. (1986). Models of Cognitive Behaviour in Nuclear Power Plant Personnel. U.S. Nuclear Regulatory Commission, NUREG-GR-4532, Washington, DC.

Woods, D.D., Roth, E.M. and Pople, H. (1987). Cognitive Environment Simulation: An Artificial Intelligence System for Human Performance Assessment, Volume 2 : Modelling Human Intention Formation. NUREG-CR-4862, Washington, DC.

Chapter 5

"NEAR MISS" REPORTING PITFALLS FOR NUCLEAR PLANTS

Geoffrey Ives
Colenco Ltd.
Baden, Switzerland

1. INTRODUCTION

The general public have become increasingly aware in recent years, through the occurrence of major accidents and complacency about modern day living, of the need to improve safety and to protect the environment. Major accidents which impact safety, productivity, availability and the environment, are rarely the result of a single error, they are more often the result of a relatively insignificant fault or error imposed on a number of standing faults. In order to improve performance therefore, it is important to reduce the existence of standing faults, and to minimise the frequency of errors.

Improvement of performance can be achieved in a number of ways, some proactive, some reactive and some by a combination of both. In particular, in the nuclear industry, performance can be improved by exploiting the opportunities offered by nuclear power plant events, potential events and near-misses. Such events provide unique opportunities for improvement of performance, by analysis, by identification of the imperfections in organisations, people, materials and activities in order to develop remedial measures for the prevention of recurrence or the development of error tolerant systems. Such opportunities may be lost forever unless they are accurately reported, rigorously investigated and expeditiously acted upon.

Near-misses, regarded as incidents which under different circumstances could have far more serious consequences, are particularly valuable because they provide analysis opportunities and timely warnings to take evasive action, without the trauma associated with the disastrous results which usually accompany major accidents. The fact that near-misses often have no tangible or observable consequences means that they are not reported, in comparison with the consequences of major accidents which are generally obvious and lead to the inevitability of reporting. There is however no ideal system of reporting, each system has to be tailor-made to the particular circumstances. A reporting system suitable for the initial start-up of a nuclear reactor may not necessarily be suitable for long term operation.

Effective reporting and analysis systems can only serve to improve performance if the information provided is acted on. Actions necessary to improve performance which get lost in the bureaucratic system or get blocked as a result of organisational conflicts may quickly lead to safety, availability and financial penalties. Reporting and analysis systems must have full management support. Unscrupulous managements who use reporting systems for purposes other than those intended can quickly destroy systems which may have taken years to develop. The use of reporting systems for disciplinary, licensing, legal, insurance or medical reasons should be avoided.

As an illustration of near-miss reporting pitfalls, three typical real-life examples are given as case studies.

2.   CASE 1: HOW TO KILL OFF NEAR-MISS REPORTING

The case in question concerned a large non-European state-controlled electricity utility, embarking on it's first nuclear plant. An expert was employed on contract to write a utility standard for reporting and processing occurrences. The plant management were required to comply with the standard, and for this purpose they wrote an implementing procedure. The standard had an event classification system, using three classes of events, classes 1, 2, and 3, of which class 1 was the most severe. In order to help the plant staff, who were embarking on a new venture, a number of examples corresponding to each of the classes was given in addition to a definition of each class.

The text of the standard contained a section about the importance of reporting near-misses (precursors to accidents). In order to encourage reporting, examples of near-misses were given, with the classification of a near-miss, in general, being one class less than the corresponding "hit". The standard required a site review, and a head-office level review of all reported occurrences. There was good cooperation between head office and the plant to get the system underway. Apparent inconsistencies between the definition of the class, and the examples were fairly quickly resolved, with the help of head office. A data-base of events was established, with an extensive "sort" facility. The system became very well respected and well used, and much improvement of performance was achieved through experience feedback. Near-misses were reported freely, from which much information was gained in order to facilitate performance improvement and the development of "error tolerant" and "error recovery" systems.

From time to time, every Company has to reorganise, - a new Chairman, productivity, profitability, financial targets, cost centres, profit centres, accountability and performance. Performance and the measurement of performance became extremely important although in the process, performance and effort often became confused as did responsibility and accountability. With such confusion, attempts to quantify various things were bound to lead to meaningless results in some cases. In order to measure performance of the plant management, (although in an interconnected system the management do not always have full control of the plant) it was decided to attempt to quantify the unquantifiable and to judge the performance of the plant (one input) on the reduction in reportable occurrences which could be achieved.

The result was immediate and spectacular. The reportable occurrences dropped by 50% in the first month and even more in the subsequent months. An analysis of the situation showed that the actual occurrences had varied little, only the numbers reported had changed. Further analysis showed that some of what had previously been reported as class 2 events were now being reported as class 3 events, and some class 3 events were not being reported at all. Near-miss reporting dried up almost completely. Wheras occurrences are almost always detectable, near-misses are often undetectable and it was clear that in the plant staff minds there seemed little point in reporting what, at the best of times, could be regarded as contentious and which would ultimately lead to their own demise. There followed time-consuming, unproductive, and at times, acrimonious disagreements between head office and site, about the correct classifications of individual occurrences. Even worse, it was realised that the near-misses not being reported at all were gone for ever.

Destroying a well functioning system by unthinking management is easy - re-establishing the system subsequently, and the restoration of confidence for using the system is difficult and takes a long time. In this particular case, it was necessary, because of internal politics, to develop a completely new system.

3.   CASE 2: IDENTIFICATION OF NEAR-MISSES

The case concerned the start-up of a new-to-type of reactor in Europe with which there had been a number of difficulties during the building, equipment installation and early commissioning. The difficulties were considered by some to be due to commencing the engineering prior to completion of the detailed design. There was intense pressure to complete the project on schedule, despite the setbacks, because it was considered to be a high prestige project. As a result of intensive shift work with long hours, in an attempt to maintain the schedule, records were not being kept properly and almost identical failures were being made repeatedly. Yesterday's near-misses became today's "hits" - sometimes there were a number of almost identical near-misses before they became "hits".

Although the early commissioning work had been impacted considerably by this method of working, none of the individual effects had been too serious because they had been limited to relatively small pieces of the plant. There were however some concerns because it was not wished to proceed with the reactor start-up in the same cavalier manner. The concerns did not extend so much to reactor safety because automatic reactor protection systems prevent serious problems by tripping the reactor if any pre-set parameters are exceeded. The concerns were orientated more towards maintenance of the programme and the avoidance of spurious reactor trips. It was acknowledged that many of the technological processes involved in reactor start-up are continuous and proceed at their own pace. From an operator viewpoint it is just not possible to stop the proceedings to write reports about what might be considered to be a near miss. Furthermore, it was realised that recognition of a near miss itself is often a very difficult thing, particularly in what, in this case, was regarded as pioneering work.

In order to attempt to gain maximum feedback from the first start-up, it was decided to relocate a number of experts into a special office immediately adjacent to the control room. When some critical phase of the operation was being undertaken, the operator would call the experts to witness the events. This would leave the operator free to concentrate his efforts on the main job of running the plant, and at the same time allow the experts to continually analyse the near-misses and potential problems.

The system worked moderately successfully. The initial difficulty concerned the inadequate number of experts and their lack of experience in operating on shift work. Some lack of expert cover resulted in a number of near-misses being lost from the data collection system and some of these inevitably resulted in incidents later on. Another difficulty, even for the experts, concerned the rapidity at which obstacles were approached which made the effectiveness of the consequent evasive action by the operator difficult to analyse. The operators readily accepted the presence of the experts, as they saw the situation as one of management support for their main function.

Even though the exercise was recognised as being only a partial success, it was none the less more successful than doing nothing, and many improvements were made as a result. The longer term benefit however resulted from the realisation that a close association between operators and experts during the operating phase could improve the design, rather than relying on operator feedback alone. The operators' problems are often immediate and pressing (akin to fire fighting) and leave little time for the formulation of longer term strategies for solutions. This led to the establishment of a "Project Development Team" who continuously sought out operating problems and engineered their solutions for the improvement of performance.

4.   CASE 3: I DON'T WANT TO KNOW ABOUT NEAR-MISSES

Traditionally, Project Managers pursue a policy of "maintenance of the aim", which for them often means the three principles of
- specification
- budget
- schedule
irrespective of a continuous stream of suggestions and requests for a change of direction. Some of the reasons for changing direction are often quite valid, but a Project Manager may find less exposure by keeping to a hard line on his original three principles and not changing at all.

An electricity utility, building their first nuclear PWR plant, had an agreement with a foreign utility (who were ahead on building a similar plant) to receive early warnings of "near-misses". A co-operation agreement with the foreign utility for the provision of early warning information cost in the region of 400,000 dollars per year, which was considered not inexpensive, but was none the less considered to be cost-effective. It was believed that significant savings would result from the early warning of problems and the consequent extra margins of time as opportunities to take evasive action.

One such early warning concerned the inadequacy of the plant fuel ponds to store sufficient spent fuel to allow continuous operation prior to fuel disposal from the site. Even though the original design may have been correct at the time, the changing world situation in relation to off-site treatment and disposal had rendered the designs invalid. The arguments for a change of design to the fuel ponds during the project phase were good, the situation for off-site fuel disposal would not get better, and a change of design after the start of operation would involve a major rebuild with all the attendant problems of shuffling spent fuel.

The Project Manager saw the design change as a major disruption to the project programme, with complications which made him unsure of the resources for coping with the situation. The two principal departments involved were Operations, who wanted the change during the project phase, and Construction, who did not want the change during the project phase. The Project Manager was reluctant to openly oppose the change, because he knew what a disruptive effect this would have on the subsequent operations programme.

The discussions went on for so long however during the project phase, with much "passive resistance" to change, that change during the project phase was no longer possible. Eventually the change had to be made during the operation phase. This necessitated a shutdown of the plant with a consequent loss of generation, extensive special provisions for the movement and storage of the irradiated fuel and the requirement for many additional human, equipment and material resources. The overall additional cost amounted to some 27 million dollars.

Even effective "near-miss" early warning systems providing timely, valuable and cost-effective information are not always acted upon.

5.   CONCLUSION

Management styles, like managers, come and go. A manager newly brought in to "clean up" a Company, may produce in the short term, highly visible financial returns for the Company and an enhanced personal reputation for the manager, but which may introduce negative longer term effects for the Company. Sweeping clean in this way may result in the good being swept away with the bad. Well established, successful, viable institutions may suffer irreparable damage which only becomes fully apparant long after those responsible have left to begin another clean up operation. It is vital that managers who indulge in Company restructuring, adopt a management style which proceeds without undue haste and with due regard for existing institutions and infrastructures. In particular, event reporting and experience feedback systems should not be used for performance appraisal, disciplinary or legal reasons.

There is no single ideal event and "near miss" reporting system. Systems should be established, developed and adapted to individual situations. Additionally a system may not be totally suitable for all phases of a project, although in the interests of consistency it may not be prudent to change a reporting system at each phase of a project. In such a case, it may be necessary to utilise one optimised system throughout the whole project in order to maximise overall benefits.

Reporting systems should be monitored throughout their entire life, especially at times of intense activity, in order to demonstrate their effectiveness. Support for reporting systems may be necessary at particular times or within particular phases so that desirable results become feasible.

Reporting system specialists should do more than stay at the work station and process reports. Such specialists should go into the field to monitor production of reports and learn first hand of the difficulties by actively assisting with report compilation to facilitate enhancement of overall performance.

Early warning of potential problems or "near misses" can be of tremendous benefit by providing the data necessary to facilitate evasive action and thereby avoid undesirable consequences. Organisations however are not always geared up to take decisions on potential problems. It is often much easier to take decisions on real problems rather than decisions on the avoidance of potential problems. In the real world however, potential problems without decisions and actions tend to become real problems.

There is a style of management which:
- prefers no decision at all because it is unable to come to terms with wrong decisions however infrequently made,
- tolerates those who say "this is unacceptable to me" without making any effort to indicate what would be acceptable,
- gives disproportionate recognition to those who criticise rather than create.

A "negative" management style of this nature, when coupled with the requirement for decision making at the upper levels of the organisation, can result in indecisiveness. Different departments often have competing or even conflicting objectives which, particularly in the case of conflict, need resolving ultimately by senior management. Decisions which appear to favour one side or the other are often difficult to make. Talking about a problem without any firm decision will perpetuate the status quo and may allow a potential problem to become a real problem. "Near miss" reporting is likely to yield more beneficial results with a decisive style of management.

Chapter 6

DEVELOPMENT OF A NEAR MISS MANAGEMENT SYSTEM AT A CHEMICAL PROCESS PLANT

Tjerk W. van der Schaaf
Department of Industrial Engineering and Management Sciences
Eindhoven University of Technology
Eindhoven, The Netherlands

In this chapter we will first briefly describe the situation at the Rotterdam Aromatics Plant (RAP) belonging to the Exxon Basic Chemicals Division, and their main reasons for starting a joint safety management project with Eindhoven University of Technology. Then the development of a Near Miss Management System (NMMS) will be outlined, emphasizing the design of a classification model for system failure. Next we will discuss organisational aspects involved in the gradual implementation of the NMMS, followed by the presentation of some preliminary results. Finally further developments in this ongoing project are mentioned, ending with an interim evaluation of the project.

1.   SAFETY MANAGEMENT AT EXXON-RAP

At the start of this research project in 1988 RAP employed approximately 160 employees, 60 of whom were process control operators and as such the "target group" at the outset. They controlled this highly automated aromatics plant by supervising the process from a Central Control Room (CCR) equipped with visual display units (so-called computer screens) as the human-machine interface. The 5 shifts of operators, at that time, had been working for about 16 years (approximately 5 million "man" hours) without a single "Lost-Time Injury" case, which meant an excellent "safety performance" by that standard.

This long period without serious injuries however did not mean that the entire system of safety control had been without any feedback concerning the more hidden problems.It was generally acknowledged that a sizable number of sofar nonconsequential errors and failures were probably happening without any systematic understanding of their causes and possible consequences. Statistically speaking, one day some of these hidden dangers could very well result in an actual injury. The existing incident- and near miss reporting system did produce some reports, but these signals originated mainly from simple, well-known hardware-related problems in the plant "outside". From the CCR, truly the information processing "heart" of the entire plant control, nothing was being reported at all.

This was considered all the more unsettling because plans for further auto-mating the plant in the future would mean that the CCR operator's task would become even more crucial to plant safety. A further concern on the managers' side was the fear that after such a long time without real injuries people would tend to become less safety-conscious during task performance.

Summarizing the above in terms of the purposes mentioned in Chapter 1, all three purposes were relevant for setting up a near miss reporting and analysis system in this case: *modelling* of the effects on safety of operator behaviour, especially of the more cognitive tasks in the CCR; *monitoring* of the real effectiveness of the existing safety management system, and increasing its efficiency by allocating its resources more optimally; and finally keeping up the *motivation* to remain alert by being regularly confronted with warnings in the form of reported near misses.

2.   DESIGN OF A NMMS

Following the reasoning of Chapter 3, a complete NMMS with all 7 modules would have to be designed. These are briefly described below, in the order of "processing" a reported near miss (see Van der Schaaf, 1990; 1991).

Module 1 (Detection) was actually a redesigned version of the existing "incident and near miss reporting form". The usual categories for pointing at "the main cause" and contributory factors were deleted, leaving the reporter to only briefly mention the near miss and suggest possible ways of improving the situation giving rise to it. Also the routing through the organisation was changed in order to minimize the response delay from the safety coordinator and the production manager. In addition a *reference database* was created by having extensive confidential interviews (see also Van der Schaaf, 1988) with 35 process operators on a recent near miss of their own "choice". In this way we hoped to be able to later evaluate the presence of any biases on the part of operators to report more freely on certain types of events than on others. Also we encouraged the interviewees to preferably report on a CCR near miss in order to gain more insight into errors and recoveries of diagnosis, commu-nication, etc.

Module 2 (Selection) is meant to act as the main point for deciding which purpose(s) shall have priority in processing a given near miss: known problems will follow the monitoring mode (see figure 2 of Chapter 3), new problems will be used for modelling, while some of the previous categories might also be used to provide detailed, convincing examples to motivate safety awareness.

Module 3 (Description) contains in the modelling and motivation modes a qualitative adapted form of a traditional fault tree, called *"Incident Produc-tion Tree"*. It contains as elements not only faults and errors, but also recoveries (see Van der Schaaf, 1988). In addition "neutral" elements may be added in order to make the description more complete: these are usually sys-tem characteristics (like the fact that in a continuous plant shifts of operators take over each other's tasks) which may be relevant to understand the sequence of events but cannot be labelled as positive or negative. The relationships between these elements may be logical (AND- and OR-gates) or chronological (indicating sequences of elements, or permanent presence of elements).

Module 4 (Classification) has as its basis mainly the well-known hierarchical "SRK" model of operator behaviour (see Rasmussen, 1986). In the Appendix to this chapter a short explanation of the Skill-, Rule- and Knowledge-based model is given. This model was chosen because of its wide acceptance by both researchers and practitioners worldwide, thus making the possibility of a comparison with other data bases more likely.

The classification consists of two phases. In the first phase a *rough classification* is made between 3 main categories: Technical factors, Organisational factors, and Behavioural factors, Each main category is, in the second phase, subdivided into several subcategories for further *detailed classification.* Each *element* of an Incident Production Tree can in principle be classified in a rough or detailed way, according to the following procedure: the first question to be answered always is whether it is a *Technical factor,* with subcategories such as: Engineering, Construction, Material Defects. If it doesn't fall in any of those subcategories, then the second question is whether the element is of an *Organisational or Management* nature: an example of a subcategory here would be the quality of operating procedures (completeness, accuracy, ergonomically correct presentation, etc.), or management pressure to let production prevail over safety.

When this second main category *also* would not be applicable, then, and only then, do we end up in the last main category, that of *Operator Behaviour*. This category has the most subcategories of all (namely 12) because we consider this category as our major point of interest (see figure 1). There are four groups of subcategories here:

K - referring to Knowledge-based behaviour,
R - referring to Rule-based behaviour,
S - referring to Skill-based behaviour, and
C - referring to physical and mental Capacities which are "present" in the operator when he enters the company grounds.

In two other chemical plants in the Netherlands this classification scheme had been "tested" previously in pilot projects with satisfactory results.

Module 5 (Computation) involved using a dBase 4 software package to store the elements and their relationships of the Incident Production Trees and their associated classification codes. Statistical analysis is handled by special software taking dBase 4 files as its input and producing its output mainly in graphical form.

Module 6 (Interpretation) is built up of links between statistical patterns of causal elements of large numbers of Incident Production Trees on one hand, and the new Exxon Safety Management Practices (SMP's) on the other hand. These SMP's together constitute a comprehensive set of possible measures to influence safety, ranging from design requirements to emergency planning. The advantage of using these global SMP's is again the improved communication possibilities with other data bases of the company.

Module 7 (Monitoring) will be handled not only by the feedback loop back to module 1 as indicated in Chapter 3, but also by an already existing independent measurement of safety performance. This "Safety Compliance Rating" index is made up of observational data and auditing inputs from a large variety of hardware, organisational and human behaviour factors.

code	description	example
K1	system status	not realising that part of the plant is inoperative because of maintenance
K2	goal	aiming at "overspec" production instead of at "right-on-spec"
R1	license (permanent)	not qualified for a certain task
R2	permit (temporary)	no permit obtained, although required
R3	coordination	not informing control-room operator of one's actions outside in the plant
R4	checks	not ensuring that system status is as expected
R5	planning	choosing wrong method for correct goal
R6	equipment/information	using wrong tools/process data
S1	controlled movement	making typing error on keyboard
S2	whole-body movement	slipping, tripping, falling
C1	permanent capacities	insufficient strength or eyesight
C2	temporary capacities	incidental use of alcohol or drugs

Figure 1: The 12 subcategories of Operator Behaviour, along with brief examples of typical errors of a plant operator's task.

(Any element which may not be unambiguously classified in any of the above (sub)categories, is coded as "X", that is "unclassifiable".)

3.   ORGANISATIONAL ASPECTS

Before specifying the implementation aspects, again in the order of modules 1 through 7, three important general aspects must be mentioned first (see Van der Schaaf, 1991):
- *management support* needed to provide the level of trust required for any voluntary reporting system; employees are guaranteed that the NMMS acts as a *learning instrument* only;
- *extensive end-user participation* in the design of all modules;
- *feedback to personnel* about all NMMS aspects: not only can the "progress" of individual reports be traced by the reporting persons, but also the NMMS output in general is quite frequent (monthly reports available to all; special near misses are mentioned in the weekly magazine, or even in instantaneous warning flyers).

Other aspects more related to specific modules are:
- the Detection module is actually a modification of an existing and well

known reporting form;
- a dozen employees have been *trained* in qualitative fault tree analysis by an external training institute to stimulate using the Description module;
- the Classification has been *supported* by implementing a question tree on a PC in the form of a series of simple yes/no questions. Numerous *help screens* explain the background of the question tree and give examples of typical causal factors for every classification category;
- when an element has been classified, it is automatically added to the database of the Computation module; graphical output provides easy overview of the results;
- as stated before, the Interpretation results are linked to *well-known labels* (the SMP's) of possible types of measures;
- the NMMS also provides extensive *administrative support* for the Monitoring module: feedback to other parties concerned can be generated automatically through the use of a Local Area Network.

4.   PRELIMINARY RESULTS

The number of near miss reports has increased by 300% to more than 10 per month since the first parts of the NMMS have been implemented (see also Van der Schaaf, 1991). More important however is the fact that now CCR near misses are also being reported and that the system has been accepted and promoted by its main user, the local safety coordinator.

Analysis of the information reported in "old" versus "new" near miss reports, and the results of classification by trained versus untrained persons has indicated the following:
- when we accept the results from the confidential interviews (see section 2) as the reference database we find that the biases in favour of reporting technical/hardware elements and against reporting organisational factors diminish with the new forms, classified by trained employees;
- within the main category of "Operator Behaviour" people tend to report and classify more Knowledge-based elements as compared to Skill- and Rule-based factors than before the NMMS was introduced.

5.   FURTHER DEVELOPMENTS

In general, the NMMS will first be completed for the process control group, and then *extended* to include also the maintenance and administrative departments. Later on also the external contractors may be integrated, as well as other Exxon plants.

An important extension will consist of building and testing *a model of (human) recovery* to function next to the existing system failure model. This may generate ideas not only for error prevention measures but also for recovery promotion.

Detection of near misses will be further supported by using a videofilm as a *training exercise for recognising* all types of factors underlying near misses. Using the new RAP process *simulator* will provide new possibilities to *generate errors and recoveries* in process control and in fault diagnosis tasks.

The reliability and validity of the Classification support software is being tested and may lead to further refinements, like better examples, questions phrased otherwise, etc. All shift supervisors will also receive *training* in the Classification module, in order to understand exactly what happens to a report sent to the safety staff, and thus to motivate their shift members to report freely and frequently.

6.   EVALUATION

Although the project is still ongoing it is worthwhile to try to (subjectively) estimate the level of progress made on each of the three purposes mentioned earlier (see Van der Schaaf, 1991):
- *modelling* has certainly been improved because of the insights into CCR task performance;
- *monitoring* may be judged soon when the database has grown sufficiently to apply the required statistical tools;
- *motivating* seems to be improved, not only with the safety staff, but also with the management and operator levels.

Now, in 1991, RAP has been working without Lost-Time-Injuries for over 18 years (6 million person hours). "Fortunately" the measurement of performance of the NMMS *on that level* is still as difficult as it was 3 years ago. The acceptance and overall correct and intensive use of near miss reporting however will have to do as probably the best performance measure available at this moment.

ACKNOWLEDGEMENTS

The contribution of Mr. L.A.A. Bollen and Mr. J.C. Koppelman to the design and implementation of NMMS so far are gratefully acknowledged.

REFERENCES

Rasmussen, J. (1986). Information processing and human-machine interaction. North-Holland, Amsterdam.
Schaaf, T.W. van der (1988). Critical incidents and human recovery: some examples of research techniques. In: L.H.J. Goossens (Ed.), Human Recovery: Proceedings of the COST A1 Working Group on Risk Analysis and Human Error, Delft University of Technology, 13 October 1987.
Schaaf, T.W. van der (1990). An integral approach to safety management. Eindhoven, University of Technology. Report EUT/BDK/42.
Schaaf, T.W. van der (1991). Implementation of a Near Miss Management System at Exxon Rotterdam Aromatics Plant. Eindhoven, University of Technology, Report EUT/BDK (in press).

APPENDIX

Rasmussen's SRK Model

The main category of Operator Behaviour in the Classification module is based on *Rasmussen's* theoretical work on the analysis of operator tasks. According to his model *three levels of operator behaviour* may be distinguished:
- *Skill-based behaviour*, referring to routine tasks, requiring little or no conscious attention during task execution. In this way enough "mental capacity" is left to perform other tasks in parallel. Example: an experienced car driver travelling a familiar route will control the vehicle on a skill-based level, enabling him/her to have an intelligent discussion, parallel to the driving task, with a passenger.
- *Rule-based behaviour*, referring to familiar procedures applied to frequent decision-making situations. A car driver integrating the known rules for right-of-way at crossings with stop signs or traffic lights, to decide whether to stop the vehicle or pass the crossing is functioning at this level *also*: the separate actions themselves (looking for other traffic, bringing the vehicle to a full stop, changing gears, etc.) will again be performed on a skill-based level. Making these familiar decisions and monitoring the execution of the skill-based actions requires some part of the total mental capacity available to the driver, but not all.
- *Knowledge-based behaviour*, referring to problem-solving activities for instance when one is confronted with new situations for which no readily available standard solutions exist: The same car driver approaching a crossing where the traffic lights have broken down during rush hour will first have to set his primary goal: does he want to proceed as fast as possible or does he want to minimise the chance of collision? Depending on this goal he will control the vehicle with varying degrees of risk taking (e.g. by ignoring some of the usual traffic rules whenever he sees an opportunity to move ahead somewhat).

Chapter 7

IDA: AN INTERACTIVE PROGRAM FOR THE COLLECTION AND PROCESSING OF ACCIDENT DATA

A.R. Hale, J. Karczewski, F. Koornneef, E. Otto
Safety Science Group, Delft University of Technology
The Netherlands

1. INTRODUCTION

The collection and analysis of accident and incident data for use in research, in decision making in companies, and as the raw material for risk assessment has always been a problem area. The difficulties involve:
- the availability of the data (the problem of underreporting),
- the quality and depth of information collected about any one incident,
- and consequently the value of the data for research or management purposes.

It is easy to be sceptical about the whole value of collecting and analysing accident and incident data simply based upon its normally poor quality. This attitude encourages a self-fulfilling prophecy: accident data is not very valuable; therefore it is not worth spending much effort on collecting it; therefore what is collected is worthless. A project is under way in the Safety Science Group in the Delft University of Technology to improve the quality of data collected on a routine basis about personal accidents and incidents and then to assess as objectively as possible if and when it can be of value. This paper describes some results of this project which has been carried out in three types of organisation:
- goods handling companies in a large harbour which were attached to a regional occupational health service (Hale et al., 1987)
- two large chemical companies
- the anaesthetics department of a teaching hospital

The first two projects were confined to injuries and damage accidents, the third has been concerned entirely with near misses.

2. PROBLEM DIAGNOSIS AND PROPOSED SOLUTIONS

2.1. Ease of data collection

2.1.1. Problem
The data normally collected for national statistics is limited in scope and depth (Smith, 1984), and subject to severe restrictions in quality and compati-

bility across countries (Glendon et al., 1986). Only a few variables are collected, mostly those which concern the injury, the injured person, the immediate cause of injury and some data on location.

The underlying causes of the accidents are rarely touched upon (Carlsson, 1984). Company statistics suffer from many of the same limitations, with the result that those who collect them are often at a loss to know how they can fruitfully use the data they have collected (Van Vliet, 1986). At the other end of the scale highly detailed information about single incidents is collected in official enquiries into serious accidents and disasters[1], in the protocols collected by some research projects (Hale, 1970; Powell et al., 1970), and by safety specialists or reporting schemes in some organisations (Rasmussen et al., 1980; Glendon and Hale, 1985). This data is seldom fully analysed partly because it is so rich and diverse and has been collected as unstructured narrative. Analysis is therefore often limited to a few incidents which are studied in very considerable depth (Turner, 1978; Reason, 1987).

A structured method of capturing information which taps some of the richness of the detailed incident reports and so goes further than the shallow analysis of routine statistics offers the promise of much more fruitful support for decisions about accident prevention and risk assessment. Some attempts have been made to collect this sort of data with the use of structured accident reporting forms (Manning and Shannon, 1979), but a major hindrance to the success of such schemes has been the problem of designing suitable reporting forms which are manageable. Because the factors leading to individual accidents are very varied, the majority of questions on a comprehensive and detailed form are not applicable to any given accident. The result is that the person filling in the form gets lost in its complexities, or becomes frustrated, bored and de-motivated by the problems of finding the boxes which do have to be filled in for any given accident.

2.1.2 Proposed Solution : IDA - an interactive computer program

We decided to tackle this problem of accident registration by developing a flexible, interactive computer program for accident data collection. The concept behind the program was that the user should be taken through a tree of questions which would present differently worded questions in relation to each accident, depending upon the information collected in answer to previous questions.

The design of the software was subject to a number of objectives:

1. The interface with the person filling in the accident information should be as user-friendly as possible so that it could be used by non-specialists with the minimum of training. No knowledge or experience of either safety or computing was assumed to be present by the end-user.

---

[1] See for example published reports of civil aviation disasters under the rules of the International Civil Aviation Organisation, the reports of the Dutch Shipping Board (Raad voor de Scheepvaart) on ship collisions and the reports of railway accidents produced by the British Railway Inspectorate.

2. The software itself should choose the question path through the question tree dependent upon the answers to earlier questions, i.e. the tree should have built-in to it the knowledge on how to go about investigating an accident systematically.
3. The structure of the question tree should be easily modifiable by a trained safety expert, to allow regular adaptation of the questions asked on the basis of changing priorities or uses.
4. Updating of information about accidents should be simple, given that all relevant information may often not be collectable at one time by one end-user.

The resulting software (IDA = Intelligent DAtabase interface or Interactive Data on Accidents) is written in Turbo Pascal and is made up of 9 modules:
1. Initialisation and set up of the system; to link fields of different types to questions, coding tables and related help screens.
2. Design and update of the question-trees; to specify the conversation scheme for the given application.
3. Input of the accident data; for the end-user to input answers to the question scheme specified in 2.
4. Documentation and correction; to verify, update or correct data and print out raw details of any accidents.
5. Manual and help; to receive information on how the programme operates or how to code given information.
6. Statistics; to tabulate data or export it in ASCII-format to more sophisticated analysis packages.
7. Output; to define standard output documents and formats (official reporting forms, annual reports etc.)
8. Options; to customise the system and store options (e.g. question trees in different languages).
9. User groups and privileges; to define which users may have access to which modules and data.
A more detailed description of the system is contained in Hale et al (1989).

The end-user comes into contact only with the modules for input and correction of data, from which he can gain access to the information on help screens and for print-out options. The end-user is locked out of all other modules to prevent any inexpert or unintended tampering. A trained safety expert (subsequently called the 'system-user') is equipped with the password to interact with the set-up modules to alter the question tree, codes, and form of print out, and to carry out statistical analysis. In this way IDA can easily be adapted to any accident registration system. Its flexibility guarantees full compatibility with already existing documents, input procedures and accident forms, which can easily be programmed as question trees.

2.2. Collection of relevant information

2.2.1. Problem

The starting point of the research was a dissatisfaction with the quality of data collected about accidents on current forms and with its suitability for decisions about deeper research needs or prevention priorities and actions. This is more than just a question of the quantity of data about any one accident; it also relates to the model of the accident process and its control which lies behind the choice of what data to collect.

The first prerequisite is that there is indeed a model of some sort, so that data collection is not driven solely by the principle: collect what is easy to collect. The model chosen must then link to the objective of the accident data collection. It is vital, therefore to state clearly what this objective is.

2.2.2 Objectives of accident data collection

Hale (1977) summarised the various purposes for accident investigation under the following headings:
1. Prevention of recurrence
2. Allocation of responsibility for causing the accident
3. Allocation of responsibility for prevention
4. Anxiety reduction (for those involved & society)
5. Curiosity (scientific and societal)

The last purpose can perhaps better be described as "hypothesis formation" or "model building" in order to advance theoretical understanding of the behaviour of systems or their elements (people, hardware).

Each purpose requires a somewhat different model underlying the data collection. For example there is little point in an organisation with purpose 1 collecting data about determinants of accidents which it cannot influence; whilst with purpose 5 it is vital to collect data about as wide a range of determinants as possible within the area of model building. Purposes 2 and 3 can be served by mono-causal classification, while the other purposes demand multi-causal modelling.

For the in-company research described here the objective was clearly the prevention of recurrence. In the hospital study, and to an extent in the harbours study, the purpose was hypothesis-forming. The other purposes will not be further discussed here. A further question is whether an accident data collection system aimed at prevention should collect the same depth of data about all accidents on a routine basis (Hale 1975). There is a trade off between the value of routine collection of very comprehensive data about all accidents, including the most trivial, and the annoyance and de-motivation this can cause because it is so time-consuming. A compromise is to specify the purpose of the routine data collection as the recognition of patterns which must then be investigated more deeply and separately and not recorded in the routine data base.

2.2.3 Solution: chosen accident model

Since IDA is a shell programme into which any question tree can be programmed, it can cater for any underlying accident model which can be phrased in terms of a tree of questions. The primary purpose of the studies described here was to uncover factors in the accidents which could be attacked to improve the level of safety in the system. The most suitable choice of model therefore seemed to be one based on tracing accidents back to factors which represent avoidable deviations from normal system functioning. We chose to combine the energy-barrier concept found in MORT (Johnson, 1980) with the concept of accidents as deviations from normal, or desired, circumstances (Kjellén, 1983; Hale and Glendon, 1987). Such a composite model encourages collection of data about the complete accident process from the initial deviation to the damage process and about the barriers and controls which failed or were absent. The latter are susceptible to far more effective and systematic intervention than the events immediately preceding the injury or damage, which have been traditionally recorded on accident forms. Such a model provides a systematic description of *what went wrong*. In order to begin the process of describing *why it went wrong* the first step in the MORT tree classifying "less than adequate" management factors was added to the tree. It was not felt to be feasible to go beyond this on a routine basis in view of the amount of work this would entail in investigating each accident.

The general structure of the question tree used in the practical projects is given in figure 1. The tree starts with some general questions about the physical circumstances at the time of the accident, about the victim, the activity and the injury and treatment. The questions then work backwards from the material cause of injury through the events which led up to it, and finally to the initial deviations from normal or desired procedures, activities and circumstances. The technical and procedural controls and barriers which failed are identified on the way. This backwards questioning order is chosen for two reasons:

- It provides an unequivocal starting point for the questioning and avoids the considerable uncertainty and individual differences which arise if an investi-gator is asked 'to begin at the beginning'. If that is asked, some will begin only seconds before the injury, others will go back hours and sometimes days. An accident is a coming together of many threads into one incident. It avoids forgetting relevant factors if they are traced back in time from their common point.
- The second essential reason lies in the need to structure a tree which asks different questions about different types of accidents. The logic of the choice of questions is much easier to construct based upon the nature of the proximal cause of the injury and the way in which the victim came into contact with the damaging energy. This question must therefore come early in the tree. For example questions about the place from which someone fell and how they came to fall are relevant to falls from heights, while questions about how toxic materials came to escape are relevant to gassing accidents (but not vice versa).

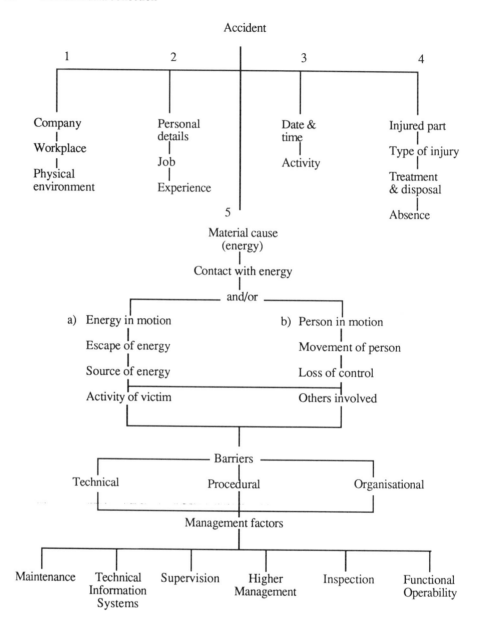

Figure 1.   Question tree structure

The disparate question paths about the different types of accident come together again at the end of tree when the absent and/or failed barriers and controls have been identified. Those questions establish in general terms what management factors led to the failures. These factors are potentially very similar for all types of technical failure. The structure of the question tree used also allows the management-oriented questions to be answered by a different person from the one who answers the simpler questions about what happened. In the harbours project the split was between ambulance personnel and safety staff; in one of the chemical companies between the accident victim plus his supervisor and a more senior departmental manager or the safety staff. The reason for this split is that the victims and personnel in the lower echelons of the organisation often do not know about the background factors to the accident and have little chance to find them out. Our experience is also that they are inclined to give biassed answers to questions about responsibility for particular procedures or their own use of them (Adams and Hartwell, 1977).

3. FIELD APPLICATIONS

IDA was developed with the help of a project in collaboration with a Regional Occupational Health Service in the goods handling sector of a large harbour. The program has been used subsequently in the chemical industry and a hospital operating theatre complex.

The projects followed a common pattern:

1. Definition of the objectives the organisation has in the registration and analysis of its accident and incident data, including the breadth of incidents to be recorded (injury, damage, near miss, etc.), the depth of investigation of different types of incident, etc.
2. Definition of the organisational structure(s) within which the reporting and analysis will take place; central and/or distributed reporting? Which information will be input by which end-user (the victim, first aider, supervisor, manager, safety staff)? What analyses at what levels? What access should be allowed to what users?
3. Programming of the question tree and the coding framework (e.g. workplaces, job titles, machinery types). This has been done upto now by the research team, but modifications will be done by the company personnel.
4. Testing of the tree with existing records of past accidents to check if all variants are covered.
5. Introduction of the system to collect data on current and future accidents.

The projects are summarised in table 1. according to the steps set out in van der Schaaf (chapter 3 of this volume).

4. DISCUSSION

None of the projects has run for long enough to make definitive statements about the achievement of their full objectives. The process of introduction of the program and the discussion which that has generated in the organisations concerned has, however, revealed a number of important factors in the likely success of the projects. These are set out below, using again the headings from van der Schaaf. There are also a number of general organisational issues which will be summarised at the end.

Table 1.   Summary of IDA projects

Step in model	Harbours	2 x Chemical works	Hospital
Detection	Injuries reported to ambulance room of occ. health service. 632 recorded	Injury + damage reported to safety system. 83 recorded in one company, ± 120 in the other	Slips/errors reported to research coordinator. ± 350 incidents
Selection	All reports	All reports	All reports
Description & Classification	Tree of what happened	Tree of what happened + elementary MORT analysis	Tree of deviations from normal activity
Computation	For all studies, cross tabulation + export facility from IDA to standard statistical packages for black spot analysis		
Interpretation	Epidemiological study + prevention recommendations	Detection of inadequacies in existing system + central assessment of trends	Hypothesis forming over possibilities for intervention in deviations
Monitoring	Projects not yet far enough advanced for this step		

4.1. Detection

In the harbour and chemical industry projects no change was made in the existing method of detecting incidents for registration. They were accepted with their existing inadequacies. These were found to be particularly troublesome in the harbours study, since reporting to the ambulance service was not obligatory; if an injured person went straight to his own doctor or to hospital, this could escape from the registration. Analysis of the recorded accidents also showed that there was a variation in the seriousness for which the ambulance was called out by different firms. These shortcomings rendered the data almost valueless for epidemiological purposes because of uncertainties about the completeness and biasing of the sample.

The problem in the chemical companies lay in the smallness of the sample of accidents in a period of a year. In both companies the detection of damage and injury incidents occurring was high. The only alternative in order to obtain larger numbers for statistical analysis would be to include near misses and errors. The experience in the hospital shows that IDA can collect data on these succesfully, but the purpose there was model building. It is an open question whether such incidents are of value for more routine prevention purposes and whether they can be collected on a routine basis; the hospital project is only for a limited duration and is known by all participants to be such.

4.2. Selection

No selection was made in the projects; all reported incidents were recorded. There were clear indications that this produced dissatisfaction and complaints of time wasting on "trivial" accidents. As a result of this the possibility was built into the question tree to either limit or extend the questions asked about particular types of accident. The greatest difficulty here is to develop criteria on which to make this choice. Ideally the incidents which should be investigated in depth are those which:
- reveal unsuspected problems in the safety system
- come from known problems which are not (yet) satisfactorily solved.

Unfortunately it is hard to know that one has such an incident until it has been fully investigated. Coupling depth of investigation to seriousness of injury is only a logical step if seriousness is lawfully related to type of cause. This is likely only to be partially so. A potential way out of this last dilemma is to ask early on in the investigation for a judgment of maximum credible injury and to use this as a filter for depth of investigation.

An additional way of handling this issue is the facility built into IDA to extend the question tree, temporarily or permanently, at any time. This means that initial pattern analysis can be used to pinpoint a type of accident which occurs more frequently and to require subsequent cases to be investigated more deeply.

A more fundamental question is whether incidents with trivial consequences, if investigated in depth, will uncover the same system weaknesses that lead to more serious consequences. Such an assumption is commonly made, based on the authority of such writers as Heinrich (1931), but little proof is available. Indeed, in the sphere of traffic accidents the evidence appears at best equivocal (Williams, 1977). This does not call into question the value of collecting near-miss data for hypothesis formation and model building purposes.

4.3 Description and classification

The concept of energy transmission and barriers as the basis of accident occurrence has proved usable with little problem, even by personnel with little or no safety knowledge. The IDA-trees as a means of establishing what went wrong have therefore been succesful in practice. This cannot be said of the barrier analyses leading to the possible management and organisational factors of relevance. Two problems arose:
- the truthfulness of the responses; e.g. reports claiming that head protection had been worn even though the head injuries which occurred would then have been impossible. This is an issue which can only be tackled if a totally different atmosphere is created, which allows for breaches of rules to be

revealed without fear.
- lack of knowledge or access to information about safety systems by those answering the questions; e.g. as to whether there was an agreed job procedure for a given task. This pointed to the need either for the tree to be filled in by several people successively, or for one investigator to collect all data before entering it in the computer.

A further problem arose in trying to program detailed trees to investigate why any given management system had failed. The potential reasons are so diverse and so specific to a given system, that they are not generically programmable. It must also be asked whether any organisation will have more than one accident in a reasonable time period which ends up in each of the large number of categories which would then be created. If this is not the case, there is no scope for analysis of patterns. This is the well known problem, that accidents are in detail unique. The skill is to find a level at which enough have something useful in common, namely that they would be prevented by the same intervention. This problem remains as the most central to be examined in the evaluation of the projects when they are completed. The interim solution chosen is to classify each accident by the absence and/or failure of a number of standard preventive systems (the MORT management factors), with the intention that clustering should be used to select management systems which require attention. The further research to see what their shortcomings are will be based on studies in addition to accident investigations.

A further issue is whether such a complicated tree structure is needed to arrive at the initial deviations. In the harbours project this was necessary, because the users were untrained in safety. In one of the chemical companies the decision has been taken to greatly simplify this part of the tree, because the users are safety specialists who are capable of arriving at the more fundamental factors without being lead by the nose to them.

4.4. Computation

It is a matter of question whether companies which have the sort of accident rate in the chemical works studied will ever have enough accidents of any one type to allow any form of statistical analysis. At the most a trend in total accident rate can be reliably plotted. Further subdivision can only be done in those companies by aggregation of data from several years, or from several sites.

Even where there are sufficient incidents recorded for statistical analysis, and where patterns can be recognised, the analysis rapidly reaches its limits of usefulnes, limited not by the accident data collected, but by the background information available. In particular none of the organisations studied has sufficient data in a convenient form to calculate accident rates useful for deciding on priorities for prevention. This would require information on time spent (or persons exposed) per activity or per category such as routine work vs. unplanned work. The indication from these studies is that improvements in such data collection and in linkage between data bases would repay investment more than further refinement in the accident data collection.

The tree form of data collection suggests analysis of incidents sharing common pathways, rather than the conventional form of cross-tabulation. The nature of the branching questions means that any given question is only relevant to a proportion of the accidents. Cross-tabulation of the whole data base therefore reveals many "missing values" and is valueless. This underlines the need for quantitative analysis to be limited to subsets of accidents which are homogeneous.

4.5. Interpretation

The IDA trees provide the link between types of injury and agent of injury and the prevention measures (presence and functioning of barriers) which could remove them. This link is lacking in most large scale data bases and makes them unsuitable for either prediction or evaluation of interventions (King, 1990). In this way there is a direct link with safety auditing information, which assesses directly the presence and functioning of safety management systems. A question which will be addressed in the evaluation of the completed projects is the extent to which the accident data fills out such auditing data. An analogy can be drawn with the developments in occupational disease prevention in the last decades. Occupational hygiene monitoring based on defined exposure limits has to a great extent replaced the reliance on detecting patent occupational disease as a means of controling disease problems.This is possible because the link between exposure and disease has been established with sufficient accuracy. If the link between the occurrence of accidents and auditable safety systems can be established in the same way, detailed accident data analysis may become superfluous for control purposes.

An interesting additional purpose for detailed narrative information has been revealed by the projects. It appears to be more convincing to managers than statistical digests. The very richness seems to help in getting over the conviction that "it could not happen here." This points to the need to be able to generate narrative from a data bank in order to convince managers that a given combination of determinants already has happened and to provide them with sufficient "aha-erlebnis" to take action. Such an experience is closely related to the hypothesis forming purpose of accident data collection; reading the narrative generates a believable picture of what has gone wrong and what might be done about it. The projects have therefore led us to the interesting hypothesis that companies with small numbers of remaining accidents may have reached the stage where the accident data can no longer be used statistically for prevention purposes; that their remaining purpose is to stimulate hypotheses, which they can never prove within the confines of that system.

4.6. Monitoring

The discussion above indicates that the value of accident data collection within organisations for prevention purposes is still a questionable issue. It also indicates that it is unlikely that a system can fulfil a useful function for long without being modified in the depth of information collected about a given type of accident. This stresses the importance of the monitoring step in the life cycle of such a system.

4.7. Organisational issues

Underlying a number of the points above are questions about the organisational anchoring of accident data recording systems. Introduction of the computerised systems in a number of the companies revealed starkly the inadequacy of the definition of responsiblility for data collection and a lack of clarity about who had access to what sort of information. A deeper investigation requires more time and necessitates someone ferreting out information which has not previously been linked regularly to accident occurrences. Information is power, and being given the task of uncovering it and feeding it into the data bank can alter the position of the investigator in the system. This can easily generate tensions and reveal a lack of trust which works against truthful answering of questions. Shifts in power or in job content because of a new skill element, such as use of computers can also lead to demands for upgrading of the job, as occurred in the harbours study with the ambulance drivers. Such tensions need to be anticipated and a strategy planned in advance to manage them.

5.   CONCLUSION

The computerisation of (part of) a task often forces a greater clarity of thinking about how (and indeed whether) it should be done. The projects described above are no exception. They have tended to support the idea that accident data can be of great value in hypothesis formation and model building, but that the use of such data for prevention purposes is much more problematical. IDA must still prove its value there. Such proof can only come from further projects with interested organisations. The structure of IDA lends itself to such further study as its shell can be filled with whatever tree is most appropriate for the organisation concerned and its problems.

6.   REFERENCES

Adams, N.L., Hartwell, N.M. (1977). Accident reporting systems: a basic problem area in industrial society. J. Occupational Psychology, 50, pp. 285-298.

Carlsson, J. (1983). How to use the Swedish Occupational Injuries Information System. Proceedings International Seminar on Occupational Accident Research. Saltsjobaden, Sweden, Elsevier.

Glendon, A.I., Hale, A.R. (1985). A study of 1700 accidents on the Youth Opportunities Programme. Manpower Services Commission, Sheffield.

Glendon, A.I., Hale, A.R., Booth, R.T., Carroll, C.H., Clarke, R.C. (1986). Occupational accidents and diseases: a review of data sources: consolidated report. European Foundation for the Improvement of Living and Working Conditions. Dublin.

Hale, A.R. (1970). Accidents during high voltage electrical switching. Report to Electricity Council, National Institute of Industrial Psychology, London.

Hale, A.R., Glendon, A.I. (1987). Individual behaviour in the control of danger. Elsevier.

Hale, A.R., Otto, E., Vroege, D., Burdorf, A. (1987). Accidents in Ports: report of a pilot study to set up an accident registration system in the Port of Rotterdam. Safety Science Group. Delft University of Technology.

Hale, A.R., Karczewski, J., Koornneef, F., Otto, E., Burdorf, L. (1989). An intelligent end-user interface for the collection and processing of accident data. Proceedings of the 6th Euredata Conference. Siena, March, Springer Verlag.

Heinrich, H.W. (1931). Industrial accident prevention. McGraw Hill, New York.

Johnson, W.G. (1980). MORT Safety Assurance System. National Safety Council, Chicago.

King, K. (1990). Artificial Intelligence and other modern information technologies: How these tools are making new accident prevention strategies possible. J. Occupational Accidents, 12, pp. 199-200.

Kjellén, U. (1983). The deviation concept in occupational accidents control: theory and method. Occupational Accident Group, Royal Technological University, Stockholm.

Manning, D.P., Shannon, H.S. (1979). Injuries to the lumbosacral region in a gearbox factory. J. Society of Occupational Medicine, 29, pp. 144-148.

Powell, P.I., Hale, M., Martin, J., Simon, M. (1970). 2000 Accidents. National Institute of Industrial Psychology, London.

Rasmussen, J. (1980). What can you learn from human error reports. In: Duncan, K.D., Gruneberg, M.M., Wallis, D.J. (Eds.), Changes in working life. Wiley.

Reason, J.T. (1987). The Chernobyl errors. Bulletin of the British Psychological Society, 40, pp. 201-206.

Smit, H.A. (1984). Epidemiology of occupational accidents. Part 2. The adequacy of existing statistics for epidemiological research into occupational accidents. (Epidemiologie van bedrijfsongevallen. Deel 2. De bruikbaarheid van bestaande statistieken voor epidemiologisch bedrijfsongevallenonderzoek). TNO-NIPG, Leiden.

Turner, B. (1978). Man-made disasters. Wykeham, London.

Vliet, L. van (1986). Preventie staat of valt bij betrouwbare informatie. (Prevention stands or falls with the reliability of information.) Maandblad voor Arbeidsomstandigheden, 62, pp. 478-482.

Williams, W.L. (1977). Validity of the traffic conflicts technique. Accident Analysis & Prevention, 13, pp. 133-145.

Chapter 8

SIGNALS PASSED AT DANGER: NEAR MISS REPORTING FROM A RAILWAY PERSPECTIVE

R.K. Taylor & D.A. Lucas
British Rail Research, Derby, U.K.
Human Reliability Associates Ltd., Wigan, U.K.

1. INTRODUCTION

The safe operation of railways depends on trains being controlled by observance of information provided by a highly reliable signalling system. The outcome of a failure to halt at a signal displaying a stop aspect will depend on many factors. The majority result in no collision and hence may be considered as being "near misses".

Unlike near misses in other situations, which often rely heavily on self reporting by the principal actors, each time a train passes a signal at danger it is automatically detected, recorded and then investigated. Signals Passed at Danger (SPAD) records provide British Rail with vital information on the state of the system and they are continuously monitored. The concern generated by such incidents has prompted several in depth studies aimed at providing a better understanding of the underlying causal factors leading to SPADs. In recent years a management information database for SPAD incidents has been set up. The present chapter provides details of the further development of this information system.

2. THE PRESENT SITUATION

A consistent rising trend of SPAD incidents has been apparent on Britain's railways since about 1982. Figure 1, normalised against total train miles, shows this trend. The rising trend is evident for all SPADs and for the subset which are reportable to the UK's Department of Transport. However, it does appear that the number of incidents which result in the serious consequences of a collision or a derailment has stabilised.

An investigation of each SPAD incident determines the person responsible. Figure 2 shows that in the majority of cases this is the driver of the train. The driver's "errors" are further classified into three main categories: Misjudgement, Misread and Disregard. The greatest increase is apparent in the instances of Misjudgement (see figure 2). It should also be noted that SPADs associated with the class of Misjudgement tend to have less serious consequences than those of the Misread and Disregard categories.

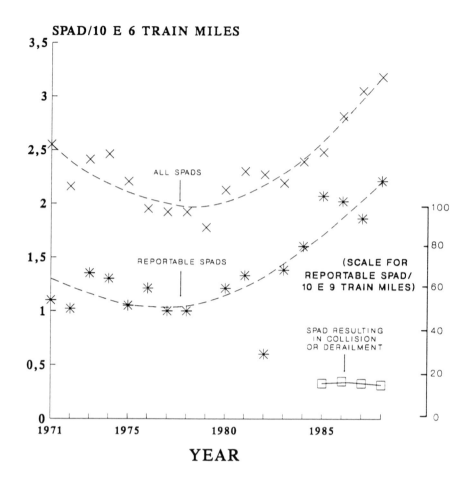

Figure 1.   Rate/mile of signals passed at danger.

3.   A EUROPEAN COMPARISON

Incidents of SPADs are not unique to Britain's railway system but recent statistics do appear to suggest that the frequency of occurrence is higher in the UK than in other European countries. Figure 3 gives comparable data for the British (BR), Dutch (NS), Belgium (SNCB), French (SNCF), West German (DB) and Danish (DSB) railways. It should be noted that the data in figure 3 is given in terms of the number of SPADS per million train miles in order to make a meaningful comparison. Table 1 gives the data for 1987 (NS data for 1984) in another form in which the raw SPAD figures are presented together with the number of SPADs per million train miles, the number of SPADs per thousand signals and the number of SPADs per thousand drivers.

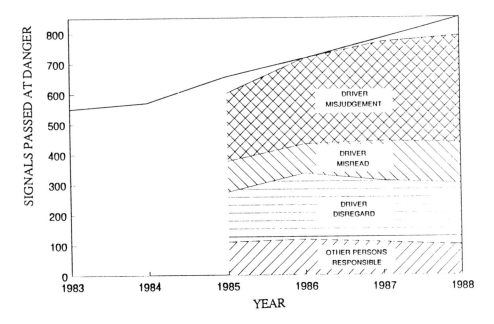

Figure 2. Contribution by error category.

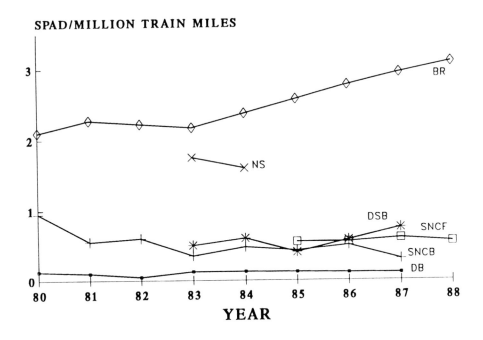

Figure 3. International comparison of SPADs for European railways.

Figure 3 and table 1 show BR at a clear disadvantage compared with each of the other continental railways where data was available. To what extent this is a true reflection of the relative performance and to what degree it is a measure of the reliability and extent of the detection and recording system is a matter of conjecture. Nevertheless, the significant difference in performance between the UK's and other railways combined with the rising trend of SPADs gave BR grounds for concern and prompted a number of initiatives aimed at reversing the trend. The differences between the different railways may be attributable to a variety of factors including: the use of lineside route signalling, a lower supervisor to driver ratio, and the widespread introduction of Automatic Train Protection (ATP) devices. It should be noted that ATP will not prevent all SPADs from occurring but should greatly reduce the number of incidents with serious consequences.

Table 1: Signals passed at danger - International comparison (1987)

Railway	SPADs	SPAD/million train miles	SPAD/thousand signals	SPAD/thousand drivers
DB	60	0.12	1.6	2.7
SNCB	18	0.30	2.1	4.5
SNCF	145	0.62	2.6	9.1
DSB	25	0.79	8.8	15.0
NS	109	1.60	16.0	31.0
BR	778	2.80	28.0	50.0

4.   PREVIOUS STUDIES OF SIGNALS PASSED AT DANGER

There have been a number of research studies into the human factors aspects of SPADs. Table 2 summarises the main factors which have been implicated in SPADs in the research to date. (This table is collated from information provided by Van der Flier and Schoonman (1988) and by Williams (1977).)

Despite such studies no clear picture has emerged of the causes of "driver error" which result in SPAD incidents. It appears that a plethora of potential causal and contributing factors must be operating. This conclusion would agree with recent studies in the wider area of "human error" (e.g. Reason and Mycielska, 1982; Reason, 1991) where both the underlying psychological mechanisms and situational influencing factors are implicated as causal factors. The implication of such findings is that any attempt to collect information into the causes of SPADs must involve a "wide trawl" for a range of human factors aspects.

Table 2: Summary of main factors implicated in SPADs in previous research

	Human factors implications of SPADs
Buck (1963) Davis (1966)	Incorrect expectation Responding to wrong signal Preoccupation and distraction Inattentiveness and timing errors
Andrews and others (1979)	Lowering of vigilance in monotonous work environments
Davis (1966)	Presence of psychiatric and/or psychosomatic complaints
Verhagen and Rijkaert (1986)	Influence of personality traits of extraversion and neuroticism
Van der Flier and Schoonman (1988)	Effect of choice reaction time behaviour Influence of job satisfaction Involvement in previous SPAD incidents
Williams (1977)	Effect of rain on distance perception
Gilchrist et al (1990)	Effect of different rolling stock Differentiation of SPAD situations Influence of knowledge, anticipation, assumptions, and misunderstanding on SPADs Role of job satisfaction

5.   PROBLEMS OF ANALYSIS

In the light of the increasing number of incidents of SPADs, British Rail decided to implement an improved Management Information and Control System with particular emphasis on the human factors aspects of such near misses. This system was to be a development of the existing reporting form and classification system. A close inspection of the existing report form revealed a number of problems.

5.1. Information on human factors aspects

The original Signal Passed at Danger form consisted of two parts: a preliminary report of circumstances, and a report on the cause and conclusions. The human factors issues covered were minimal and consisted of only the following two factors:

a. Some indication of the experience of the driver (in terms of number of years experience and age).
b. An indication of the driver's experience on this particular route (in terms of the date the route was last signed for and the number of weeks since the route was last worked).

Information on the following contributory factors was also collected:

a. Whether the driver had been involved in any SPADs in the previous 10 years.
b. Whether the weather, visibility, rail conditions or lighting conditions were implicated.
c. Whether the driver was tired. (This was indicated by the elapsed time since the last rest period.)
d. Whether the driver was distracted by the presence of another man in the cab (as indicated by the manning level and the number of people present in the cab).
e. Whether the SPAD was caused or influenced by faulty equipment.

5.2. The classification system

A driver's error causing a SPAD was classified as being one of the following categories of error:

Misjudgement

Over-running a stop signal in the Danger position, although the signal had been observed and an endeavour made to obey.

Misread

Accepting a signal in the "OFF" position which was not applicable to the line on which the train was travelling and thus passing the relevant signal in the Danger position.

Disregard

Failure to observe and/or obey a stop signal in the Danger position.

These three categories had some bearing on the causes of driver error but, by themselves, did not enable a clear, complete and unambiguous classification of the nature of the error.

5.3. Benefits of keeping the existing classification framework

This categorisation scheme had been used for some time within BR and was very familiar to future users of any revised data collection system. Existing data on SPADs could be analysed into 3 groups based on this scheme. There was therefore a strong case for retaining the use of these terms in the new SPAD data collection system for the following reasons:

a. Traction inspectors (the users) would accept the revised form more readily if they perceived it to be a development of the existing form rather than a radical departure.
b. The choice of corrective action tended to be influenced by the category of error. It was therefore highly probable that these categories would continue to be used by inspectors even if they did not appear on the form.
c. Existing data and new data would be compatible in certain aspects if the categories were maintained.

5.4. The problems of the original classification framework

However, there were certain problems with this original tripartite error categorisation scheme. In particular, the category "disregard" was essentially a "catch-all" grouping which covered at least two error types: ignorance or violation of rules and regulations, and failures to observe the signal for any number of reasons. This was probably the reason why, statistically, a higher proportion of cases of "disregard" occurred in those SPADs which had more serious consequences. In addition, there were stronger connotations between "disregard" and the concept of "wrongdoing" or culpability than with the other two terms. This may have led some inspectors to equate the causal "disregard" category with the consequences of the SPAD.

5.5. The need for a category of "misunderstanding"

One final problem with this existing scheme was that it did not explicitly address misunderstandings or miscommunication between BR staff. Previous research (e.g. Williams, 1977; Van der Flier and Schoonman, 1988) found that a small proportion of SPADs (about 7-10%) were caused by miscommunications of one form or another. There is therefore some value in separating out such instances as a distinct causal category.

5.6. Corrective actions

The corrective actions which could be taken for a SPAD were restricted in the existing form to a verbal caution, a reprimand, suspension for a variable period of time, and more severe action. No allowance was made for other useful control actions such as retraining the BR staff involved, or making modifications to equipment or working practices.

6. AN IMPROVED DATA COLLECTION SYSTEM

A new data collection system for SPADs was developed taking into account the problems and the positive features which had been noted for the existing system. The development and implementation of the new Management Information and Control System involved:

a. A revised categorisation scheme for classifying SPADs caused by driver error.
b. An extended set of causal and contributing factors for human errors.
c. A modified and extended data collection form designed according to ergonomic guidelines.
d. The introduction of additional possible corrective actions.

6.1. The revised categorisation scheme.

The revised classification system attempted to avoid the problems with the existing system particular in relation to the category of "disregard". This category name had to remain but steps were taken to prevent it from being a "catch-all" class. An additional category of "miscommunication" was included as a separate item within the scheme. This was further subdivided to enable identification of different aspects of such errors.

Table 3: Revised categorisation scheme for errors leading to SPADs

---

A. MISJUDGE

A1. Misjudgement: misjudged train behaviour
A2. Misjudgement: misjudged environmental conditions

B. MISREAD

B1. Misread: viewed wrong signal
B2. Misread: viewed correct signal, misread signal aspect

C. DISREGARD

C1. Disregard: anticipation of signal clearance
C2. Disregard: failure to check signal aspect
C3. Disregard: failure to locate signal
C4. Disregard: ignorance of rules/instructions
C5. Disregard: violation of rules/instructions

D. MISCOMMUNICATION BETWEEN STAFF

D1. Miscommunication: wrong information given
D2. Miscommunication: ambiguous/incomplete information given
D3. Miscommunication: information not given
D4. Miscommunication: correct information given but misunderstood

The rationale used throughout the revision of the categorisation scheme was to subdivide the four major categories of error in order to allow more specific identification of the causes of SPADs. The new scheme is shown in table 3. Using this new classification provides 13 subcategories which could be used to identify the nature of a driver's error more specifically. A decision tree which showed the derivation of the subcategories for disregard, misjudge and misread, together with an indication of the need for the category miscommunication, was also developed (see figure 4). It was intended that this decision tree could be used to assist in the identification of subcategories if required.

6.2. Probable causal and contributory factors

Using the available information sources a list of causal and contributory factors for SPADs was drawn up. This list was then divided into:

a. General items which should be asked for every type of error, for example,whether the driver was distracted or preoccupied, whether he had been involved in other SPADs, etc.
b. Specific items which were only relevant for a particular type of error, for example, asking about practical train handling skills is only appropriate if a misjudgement has occurred, whether the signal sighting is confusing is relevant to an error of misreading, etc.

This division meant that parts of the revised form inquiring about human factors issues fell naturally into distinct sections. Thus certain sections of the revised form cover specific issues whilst others encompass more general information.

6.3. Determining corrective actions

In the revised form the choice of possible corrective actions for SPADs has been extended considerably. The set now covers:

a. Disciplinary actions (as in the existing form)
b. Retraining/training
c. Modifications to equipment
d. Modifications to work practices (e.g. patterns of supervision, changes to shift working, delivery methods for information of route changes to drivers)

The linkage between the causes of driver error and certain of these corrective actions is quite clear. For example, retraining would be appropriate if inexperience was a major causal factor in a SPAD, disciplinary action would be appropriate in the case of a clear violation of rules and regulations.

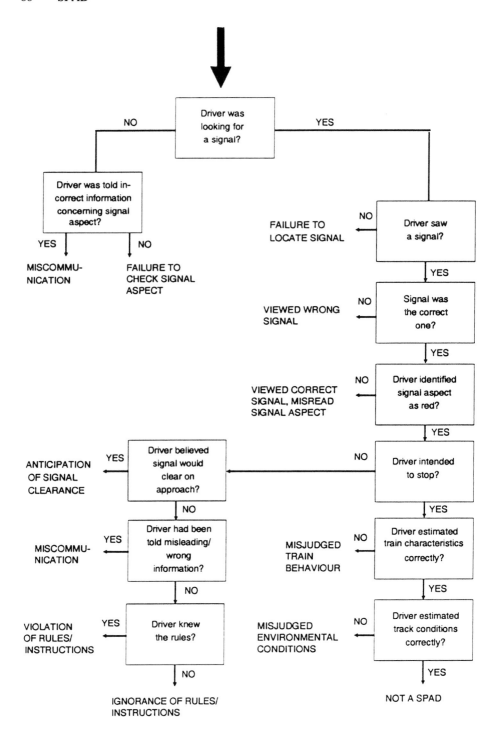

Figure 4.   Decision tree to determine category of driver error causing non-technical SPAD.

6.4 Format changes

There were some radical changes made to the data collection forms. Wherever possible ergonomic guidelines on the design of questions and forms were used (e.g. Wright and Barnard, 1975; Barnard et al, 1979). The major modifications were as follows:

a. The form is now in three parts. Part 1 is a preliminary (factual) report. Part 2 is a report on the causes whilst Part 3 forms the conclusions.
b. Each part contains distinct sections in which related questions are grouped. In Part 1 the sections cover: general details, consequences, train, braking, signalling, external circumstances, cab circumstances, and driver's details. In Part 2 the sections are related to the type of error category identified and to the allocation of responsibility for the incident between the driver, other BR staff and equipment failures.
c. Indication of who is to fill in the questions is made through the use of symbols.
d. Notes are provided to assist the form user. These are compiled as a separate document and brief cross references are given in each part of the form.
e. Two types of question predominate: simple Yes/No options and multiple choice questions. For both of these the user is required merely to tick the appropriate box.
f. Written (textual) answers have been restricted wherever possible.
g. For certain questions where the issue is not clearcut the user has been provided with a "maybe" option i.e. yes/maybe/no. This is particularly the case in Part 2 (causal analysis) where human factors considerations predominate. The rationale is that much potentially valuable information would be lost without this option.
h. Questions have been made as short as possible and they only ask about one aspect at a time.

7. IMPLEMENTATION ISSUES

In order to avoid possible implementation problems a group of users was consulted about the design of the draft SPAD enquiry forms. In addition, introductory training was delivered to users of the new data collection system and feedback from such users was encouraged.

7.1. Involvement of a user group

A representative user group was asked to comment on the draft forms and to suggest changes. A number of amendments were made as a result of this involvement. These changes were related to the terminology used and to the ordering of information within the forms. The changes were designed to reflect how the form would be used in practice.

7.2. Introduction of forms to users

A BR representative who had been involved closely with the design of the new SPAD data collection system spent time introducing users (Traction Inspectors) to the new forms and their underlying rationale.

7.3. Encouragement of feedback from users

The revised forms are designed to be easy to complete and unambiguous. However, only field experience would reveal if they fulfil these objectives. A feedback form was designed and distributed to the form users. The results of this feedback showed that the revised form was well received despite requiring three times as much information as the previous SPAD enquiry form.

8.   DISCUSSION

The new SPAD enquiry form was introduced in January 1989 and was well received by its intended users. It has remained in use since that date with only minor modifications. In parallel a new computer database has been developed to cope with the extended information collected with the revised form.

8.1. Why was the new SPAD enquiry form so successful?

Whilst it is difficult to pinpoint why the new form was so successful certain aspects of its design and introduction do appear to be important. Firstly, the form was designed around an existing SPAD classification scheme which, whilst not ideal in many respects, was known and used by Traction Inspectors. Secondly, the SPAD enquiry form was designed to be logical and easy to answer. Thirdly, a representative user group was involved in the later stages of design. Fourthly, all users were formally introduced to the system through a series of presentations. Finally, users opinions and difficulties with the new form were requested and acted upon. One overriding factor may have been the "ownership" of the SPAD data collection system which was (and still is) firmly held by the Operations Department of BR. Traction Inspectors are members of this department. Whilst the new SPAD data collection tool was designed by a research department, once completed the Operations Department once again took over ownership of the scheme.

8.2. Analysis of the human factors data

The major benefit of the new system will be the enrichment of data analysis on human factors aspects of SPADs. With the extended information content it will be possible to analyse the data at a number of levels. For example:

a. Analysis at the category level (misjudge, misread, disregard, miscommunication). This type of analysis is already performed with the existing data
   base and it is necessary in order to be able to compare the newly gathered

information with the existing data base. However, analysis at this level is of very little use in determining the cause of the error.

b. Analysis at the subcategory level. It is easier to discover the causes of driver error by examining the data at this level. The 13 subcategories (see table 3) may be used to determine the relative incidence of different error categories.

c. The impact of general factors such as preoccupation, distraction, inexperience, etc. can be established, as may the role of specific causal factors.

d. A search for predictable patterns of "high risk" combinations of factors should be made at the subcategory level. These patterns may be determined empirically or theoretically. Examples of the latter would be the search for "habit intrusions" where the combination of, for example, experience of the route, preoccupation, over-reliance on past experience with a signal and anticipation of signal clearance would all be relevant. Another example would a "rule-based mistake" indicated by the combination of unfamiliarity with the signalling layout, viewing of the correct signal but misreading the signal aspect, in the absence of any evidence of preoccupation or distraction.

8.3. SPAD reporting within the near miss reporting framework

Van der Schaaf (chapter 3 of this volume) lists seven aspects of the design of near miss reporting systems. Table 4 shows how these aspects were involved in the redesign of the SPAD data collection system.

Table 4: Relevant aspects of the SPAD near miss reporting system

Feature	Comment
1. Detection	Occurs automatically
2. Selection	All incidents are investigated
3. Description	The revised form contains more human factors aspects
4. Classification	Existing and accepted system was revised
5. Computation	A larger database and more computer power provided
6. Interpretation	Multiple levels of analysis now possible
7. Monitoring	Feedback from users sought

In addition to these seven aspects we would consider the following issues to have been vital in ensuring the success of the new form:

- relationship to the previous system,
- formal introduction of the system to users,
- involvement of users in design,

- the actual design of the form itself,
- the ownership of the scheme.

REFERENCES

Barnard, P.J., Wright, P. and Wilcox, P. (1979) Effects of response instructions and question style on the case of completing forms. Journal of Occupational Psychology, 52, 209-226.

Gilchrist, A.O. et al. (1990) An investigation into the causation of signals passed at danger.Report ref: TM TAG 138. Unpublished report British Rail Research, Derby.

Reason, J. and Mycielska, K. (1982) Absentminded? The psychology of mental lapses and everyday errors. Englewood Cliffs, N.J.: Prentice Hall.

Reason, J. (1991) Human Error. Cambridge: Cambridge University Press.

Van der Flier, H. and Schoonman, W. (1988) Railway signals passed at danger. Applied Ergonomics, 19 (2), 135-141

Williams, J.C. (1977) Railway signals passed at danger - some further research. Proceedings of the Annual Conference of the Ergonomics Society.

Wright, P. and Barnard, P. (1975) Just fill in this form - a review for designers. Applied Ergonomics, 6(4), 213-220.

Chapter 9

VIDEO ANALYSIS OF ROAD USER BEHAVIOUR AT INTERSECTIONS

Richard van der Horst
TNO Institute for Perception
Soesterberg, The Netherlands

ABSTRACT

Although accident statistics have an important general monitoring function and may form a basis for detecting specific traffic safety problems, the information available from accident data is inadequate for the next steps in studying safety problems, viz. analysing and diagnosing, defining remedial measures, and evaluating effects. Systematic observations of road user behaviour, combined with knowledge of human information processing capabilities and limitations, offer wider perspectives in understanding the causes of safety problems. With such an approach it is important to identify normal behaviour in order to be able to distinguish *critical* from *normal* behaviour. A video-based observation and analysis method was developed that enables an objective quantification of dynamic characteristics of road user behaviour in various traffic situations.

The results of several studies point to the direct use of time-related measures in road users' decision-making in traffic. Examples of these measures are the Time-To-Collision (TTC) for the severity scaling of traffic conflicts and the Time-To-Intersection (TTI) for approaching and negotiating intersections. Some examples of recent research are discussed from which it can be concluded that road users display rather consistent behavioural patterns dependent on the type of priority regulation, and that normal and critical behaviour are distinguished well by applying a minimum value of 1.5 s as a criterion.

1. INTRODUCTION

In order to answer the question how to prevent accidents a good understanding of the causation process leading to accidents is needed. However, the pre-crash phase of accidents can almost never be investigated by direct observation. Reconstructions based on laborious in-depth and on-the-spot accident investigations can give some insight into the causes of accidents, but it appears extremely difficult to quantify and to translate findings into recommendations for countermeasures related to road user, vehicle, road and/or traffic environment (Grayson and Hakkert, 1987).

Systematic observations of road user behaviour in various traffic situations combined with knowledge of human information processing capabilities and limitations may offer wider perspectives in understanding the causes of safety problems. In particular, the study of conflict behaviour is a natural candidate for that purpose; the processes that result in near-accidents or traffic conflicts have much in common with the processes preceding actual collisions (Hydén, 1987), only the final outcome is different, the frequency of occurrence of near-accidents is relatively high, and they offer a rich information source on causal relationships. The preceding process can be systematically observed, which is essential for analysing, diagnosing, and solving traffic safety problems. In this approach traffic situations are ranked along a continuum of events ranging from normal situations, via conflicts, to actual collisions. Recently, Hydén (1987) has introduced a pyramidal representation of this continuum (Fig. 1), clearly visualising the relative rate of occurrence of the different events.

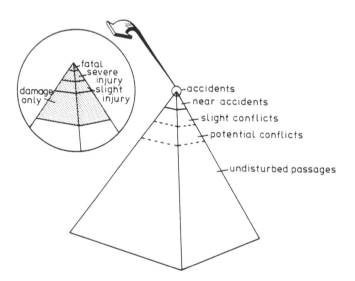

Figure 1.   The continuum of traffic events from undisturbed passages to fatal accidents (Hydén, 1987).

The analysis of road user behaviour in critical encounters may not only offer a better understanding of the processes that ultimately result in accidents, but, perhaps even more important and efficient in the long run, also provide us with knowledge on road users' abilities to restore a critical situation to a controllable one. An important issue is how to distinguish the various normal to critical situations systematically and reliably. In particular, this chapter will focus on the distinction between normal and critical behaviour of road users approaching and negotiating intersections. This approach is based on a quantitative analysis in terms of time measures such as Time-To-Collision (TTC) and Time-To-Intersection (TTI).

2. METHOD

2.1. VIDeo Analysis of Road Traffic Scenes: VIDARTS

Methods for studying road user behaviour range from simple observations in actual traffic situations, via the use of instrumented vehicles, to highly controlled laboratory experiments. Which one it is best to use is, among other things, dependent on the level of detail that is required for specific research questions. For interactions between road users or between road user and environment the analysis of road user behaviour in terms of the resulting movements of the vehicles is often appropriate. These can be observed unobtrusively from outside the vehicle in the natural setting of actual traffic situations. A video-based observation- and analysis- method was developed that enables an objective quantification of dynamic characteristics of road user behaviour. In summary, the procedure consists of video recordings with one or more fixed cameras on the spot and subsequently an off-line quantitative analysis in the laboratory with specially developed video analysis equipment. The quantitative analysis consists of selecting positions of points on a vehicle by positioning an electronic cursor on a video still. By a transformation based on at least four reference points the x- and y-coordinates of the video plane are translated into positions on the road plane. Since movements of several vehicles have to be analysed in relation to each other or to the environment, a road-based coordinate system is used. By analysing successive video stills a sequence of positions over time is obtained, from which other variables such as velocity, acceleration, heading angle, and time measures such as TTC and TTI can be derived. For further details the reader is referred to Van der Horst (1990) and Van der Horst and Godthelp (1989).

2.2. Time-related measures

Time-To-Collision
In research on Traffic Conflicts Techniques Hayward (1972) initiated a search for objective measures to describe the danger of a conflict situation and concluded that the Time-To-Collision (TTC) measure is a dominant one. He defined TTC as: "The time required for two vehicles to collide if they continue at their present speed and on the same path". TTC at the onset of

braking, TTC$_{br}$, represents the available manoeuvring space at the moment an evasive action starts. The minimum TTC (TTC$_{min}$) reached during the approach of two vehicles on a collision course is taken as an indicator for the severity of an encounter. In principle, the lower the TTC$_{min}$ is, the higher the risk of a collision will be. To illustrate TTC, Fig. 2 shows what happens when a car approaches a stationary object.

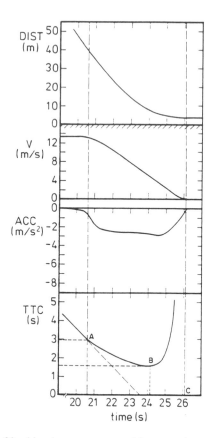

Figure 2.   Time histories of braking by a car approaching a stationary object; DIST = distance to object, V = velocity, ACC = acceleration, and TTC = Time-To-Collision based on constancy of speed and heading angle. Point A indicates TTC$_{br}$ and point B TTC$_{min}$.

Usually the concave shape of the TTC curves does not show up so nicely, since in more complex interactions of two moving road users the collision course is often ended before point B is reached. But even then TTC$_{min}$ indicates how imminent an actual collision has been. Details of the calculation of TTC can be found in Van der Horst (1990).

Time-To-Intersection
Whereas the TTC measure deals with interactions between two road users, the Time-To-Intersection (TTI) is a time-based measure to describe road user

behaviour relative to the road environment itself. For example, when approaching an intersection, TTI is defined as the time that is left before the intersection area will be entered, given by the distance to that area divided by the instantaneous speed. Fig. 3 gives some examples of TTI curves as a function of the running time t for five hypothetical but realistic types of approaches.

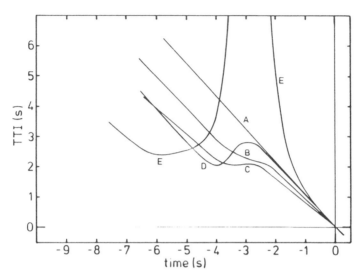

Figure 3.   Hypothetical TTI curves as a function of the running time.

For each approach the moment of entering the intersection area is taken as t = 0 s. Curve A illustrates an approach with a constant speed; TTI decreases linearly with time. By decelerating differentially it is possible to reduce the decrease of TTI (curve B), to keep TTI constant for a while (curve C) or even to increase TTI (curve D). A complete stop would result in curve E. Behaviour such as that from curve C or D makes it possible to gain time for interpreting the situation before actually entering the intersection. Similar to TTC, TTI at the onset of braking (TTI_{br}) and the minimum TTI (TTI_{min}) reached during the approach, if any, can be distinguished.

2.3. Field study

Under a contract from the Transportation and Traffic Engineering Division of the Dutch Ministry of Transport an explorative study was conducted to identify and describe the behavioural rules applied by road users in a priority situation at non-signalised intersections. In the Netherlands a general give way to the right rule applies at intersections without specific regulation, with one exception, viz. that a bicyclist or moped rider always has to yield to motor vehicles on the intersecting road. Behaviour was observed by video registration at two locations at Soest, one having a special priority regulation (indicated by yield signs and triangular road surface markings), the other without specific regulation (so that the general regime of giving way to the right with the exception for bicyclists applied). The selection of the locations was such that both had a comparable standard geometry with a distinguishable major and minor road. Also the intersection sight distances and traffic volumes (about 300 veh/h and 30 bicycles/h on the major road and about 100 veh/h on the minor road) were comparable for both intersections.

At each intersection video recordings with three cameras were made during three working days (8.00-18.00 h). Two cameras were at a high position in an adjacent apartment building, one covering the approach area of the major road and the other that of the minor road with the intersection area partly in common. A third camera was at eye-height at street level, un-obtrusively mounted in the back of a parked car and registering head movements of road users at one of the approaches.

Encounters were selected from video with a proximity criterion up to about 5 s and no more than two road users involved in the interaction. Only encounters between a car from the minor road and a car or bicycle going straight on or turning left on the major road were considered. Complex encounters with more than two road users involved were not included in the analysis. For the observation period a total of 644 encounters were selected, subdivided by type of encounter, priority regulation, and manoeuvre according to Table I.

Table I. Number of observed approaches by minor road card, both free-moving and involved in an encounter with a car or bicycle on the major road, by type of manoeuvre and priority regulation.

	left turn			straight on			right turn		
	free	car	bicycle	free	car	bicycle	free	car	bicycle
yield	23	55	17	30	193	48	30	109	35
general rule	30	43	38	30	40	28	28	19	19

In order to separate behaviour in an interaction from that conducted to negotiate the intersection per se, a total number of 171 free-moving cars from the minor road without the presence of traffic on the major road was selected for further analysis.

The original analysis consisted of the determination, by means of VIDARTS (sampling time 0.24 s), of the initial configuration and the final outcome of an encounter in terms of who was going first, the computation of positions, speeds and accelerations of the vehicles involved as a function of the running time, and the registration of when there were head movements of one of the road users to the left and to the right. For more details about this study the reader is referred to Janssen and Van der Horst (1988). In the context of the present study the data were reanalysed with respect to TTI and TTC. For both vehicles involved in an encounter, as well as for free-moving vehicles, TTI curves were calculated by dividing the instantaneous distance to the intersection area (defined as the distance to a given reference line dependent on the approach direction, see Fig. 4) by the instantaneous speed. In general, differences between conditions were tested by an analysis of variance (ANOVA) on TTI and TTC values. For further details on the statistical tests, the reader is referred to Van der Horst (1990).

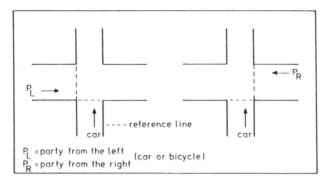

Figure 4.   Definitions of reference lines for calculating TTI dependent on approach direction (P_L, P_R = party on major road is approaching from the left or right, respectively).

3. RESULTS

At the yield intersection in total 457 encounters with two road users involved were selected, whereas at the general rule intersection only 187 encounters of this type occurred. At the latter far fewer encounters among more than two road users occurred than at the yield intersection. Since the selection of both intersections was, among other things, based on similar traffic volumes and manoeuvre patterns, the big difference in the number of encounters is remarkable. Janssen and Van der Horst (1988) give the plausible explanation that a yield regulation in fact "collects" more encounters because road users at the minor road more frequently decide to slow down when they are initially ahead of road users on the main road.

3.1. Time-To-Intersection (TTI)

Time-To-Intersection at the onset of braking (TTI$_{br}$)

Because the moment of braking could not be deduced in the video-recordings from the onset of the brake lights, a certain deceleration rate had to be set as a criterion for the moment of braking. To define this moment a deceleration level of -1 m/s^2 is taken, which generally occurs when releasing the throttle. Unfortunately, at the general rule intersection, this level of deceleration was already exceeded in about 50% of the minor road approaches at the moment of entering the video picture, while on the other hand 30% did not exceed -1 m/s^2 at all before entering the intersection area. Therefore, only the yield intersection TTI$_{br}$ data were analysed in more detail. Also the distance to the intersection and speed at the onset of braking, $DIST_{br}$ and V_{br} respectively, are taken into account.

In encounters with traffic on the main road, neither TTI_{br}, $DIST_{br}$ nor V_{br} of minor road cars differ significantly by type of manoeuvre, type or approach direction of the main road party, see Fig. 5.

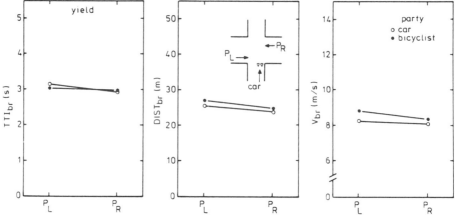

Figure 5.   Mean TTI_{br}, $DIST_{br}$ and V_{br} at the onset of braking by minor road cars (left turn and straight on manoeuvre combined) by type of party on the main road and by party's approach direction at the yield intersection.

By comparing the data of encounters with that of free-moving cars in an analysis of variance (ANOVA), however, it appears that free-moving car drivers from the minor road display a somewhat lower average TTI_{br} of 2.8 s than in an encounter, due to a higher speed. The presence of another party on the main road does not effect the distance at which TTI_{br} occurs. A comparison of the cumulative distributions of TTI_{br}, $DIST_{br}$, and V_{br} with and without a party on the main road reveals similar results (Van der Horst, 1990).

So far, the variables TTI_{br}, $DIST_{br}$, and V_{br} have been presented separately, but, of course, they are related to each other. It seems reasonable to assume that the speed with which a driver is approaching an intersection can be regarded as the independent variable. Then, given a certain speed, the decision to start braking might be either based on the distance or the time to the intersection. $DIST_{br}$ appears to be highly correlated with V_{br}, whereas TTI_{br} is not, see Table II.

Table II. Correlations and regression coefficients of relations $DIST_{br}$ - V_{br} and TTI_{br} - V_{br}, respectively. A: decelerating minor road cars involved in an encounter; B: free-moving minor road cars; C: decelerating main road cars involved in an encounter.

	n	\multicolumn{4}{c}{$DIST_{br}$}	\multicolumn{4}{c}{TTI_{br}}
		r	p	intercept	slope	r	p	intercept	slope
A	262	0.83	0.000	1.49	2.86	-0.09	0.15	3.24	-0.02
B	64	0.76	0.000	-2.24	3.06	0.13	0.29	2.46	0.04
C	46	0.61	0.000	1.02	2.30	0.19	0.21	1.82	0.06

Minimum Time-To-Intersection (TTI_{min})

If a driver decelerates to such an extent that the TTI curve has a minimum or an inflection point before actually entering the intersection area, he gains time for interpreting the situation and, if necessary, he has ample opportunity to come to a complete stop. A log-linear model analysis on the number of approaches by minor road cars with and without a TTI_{min} reveals that at the yield intersection neither the type of manoeuvre (left turn or straight on) nor the type of main road party (car or bicycle) or its direction (from the left or right) is relevant to the proportion of approaches with a TTI_{min}; a model with a constant proportion of 75% satisfactorily describes all conditions (chi^2 = 3.51, df = 11, Q = 0.982) (see Fig. 6).

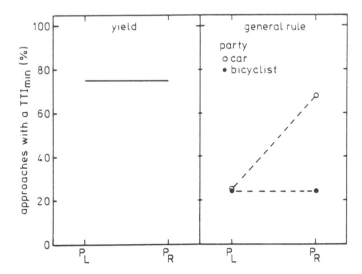

Figure 6. Results of a log-linear model fit on the proportions of TTI_{min} approaches by minor road cars (left turn and straight on manoeuvre combined) at the yield (left) and general rule (right) intersection by type of party on the major road and its approach direction.

At the general rule intersection the interaction between type of party and the direction from which the party approaches, is needed to get a good fit (chi^2 =

9.28, df = 6, Q = 0.158). The proportion of approaches with a TTI_{min} properly reflects whether one has the right of way, viz. giving the right of way to cars from the right, and having the right of way over cars from the left and over bicyclists irrespective of their approach direction. At both intersections a TTI_{min} is more frequently found in an encounter than in an approach without other traffic. The right turn manoeuvre has the lowest proportion of approaches with a TTI_{min} (Van der Horst, 1990).

At the yield intersection the behaviour of minor road drivers in encounters with traffic on the main road appears to be rather consistent with respect to the value of TTI_{min}; ANOVAs reveal that there are no effects of type of manoeuvre, the type of road user involved, or the direction the party is coming from, see Fig. 7. Less than 10% of the approaches with a TTI_{min} have a TTI_{min} less than 1.5 s. Without other traffic on the main road, the mean TTI_{min} (1.9 s) is lower than in encounters with other road users, but even then less than 6% of the approaches with a TTI_{min} have a $TTI_{min} < 1.5$ s.

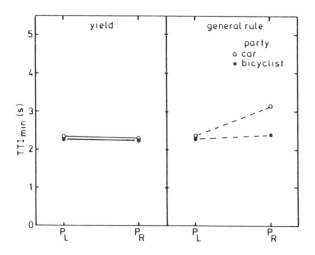

Figure 7. Mean TTI_{min} for cars approaching from the minor road (left turn and straight on manoeuvre combined) at the yield (left) and general rule (right) intersection by type of party on the major road and its approach direction.

At the general rule intersection, the behaviour of 'minor' road car drivers with respect to TTI_{min} is dependent on the right of way. If they have the right of way similar mean TTI_{min} values occur as at the yield intersection, but in encounters with cars from the right the mean TTI_{min} appears to be higher (t-test, two-tailed, t = 2.16, df = 33, p<0.05), see Fig. 7. At the general rule intersection free moving cars hardly ever displayed a minimum TTI before entering the intersection area.

Time-To-Intersection at the moment of first head movement (TTI_{1st})
The direction of the first head movement might be an indicator of the initial priority in directing attention when approaching an intersection. Fig. 8 gives

the proportions of minor road approaches where the first head movement is to the right without (free-moving) and with a party on the main road (subdivided by the direction the party is coming from). The data for car-car and car-bicyclist encounters were combined since the type of party appeared not to be a relevant factor.

A log-linear model fit reveals that at the general rule intersection the first head movement is more frequently to the right than at the yield intersection (chi^2 =12.42, df = 4, Q = 0.01). Adding the type of approach (free-moving, P_L, or P_R) also significantly contributes to the model (chi^2 = 10.24, df = 2, Q = 0.006). A separate analysis of the data at the yield intersection reveals that neither the presence, the direction, nor the type of the party on the main road influences the direction of the first head movement by minor road car drivers; a fixed proportion of 74% of the drivers first looks to the left (chi^2 = 3.28, df = 3, Q = 0.35). At the general rule intersection, however, the presence of a party from the right (irrespective of whether it is a car or a bicyclist) significantly increases the number of first head movements to the right (chi^2 = 6,84, df = 1, Q = 0.009).

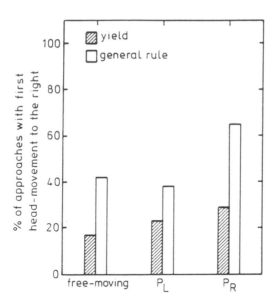

Figure 8. Percentages of minor road car approaches with the first head movements to the right without (free-moving) and with a party on the major road by type of priority regulation.

The TTI value is determined at the moment that the first head movement of minor road car drivers was in the direction of the party on the major road. This is indicated by TTI_{1st}; $DIST_{1st}$ and V_{1st}, were also determined at this moment. In Fig. 9 the mean values are contrasted with the values for free-moving cars for the corresponding directions.

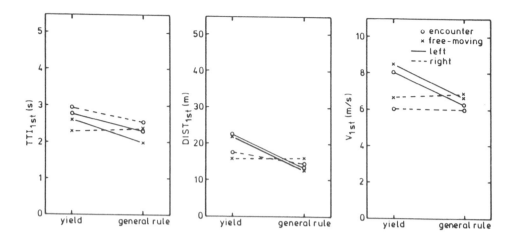

Figure 9.   Effects of priority regulation on TTI, DIST, and V at the moment the first head movement is in the direction of the party (left or right).

No distinction between car-car and car-bicyclist encounters is made since none of the variables significantly differed with respect to the type of party involved. ANOVAs on each of the variables reveal that in an encounter the first head movement is made at the same distance but at a lower speed than for free-moving cars, resulting in a higher TTI_{1st}. When the first head movement at the yield intersection is to the left, this head movement occurs at a greater distance and with a higher speed than the same movement at the general rule intersection, also resulting in a somewhat higher TTI_{1st}.

3.2. Time-To-Collision (TTC)

In the previous section the time measure TTI was used to describe road user behaviour at intersections relative to the road environment. To analyse the interactions between road users themselves, Time-To-Collision (TTC) is used if a collision course exists. The latter is the case when two vehicles will meet both in space and in time if they continue their current speed and heading. In particular, the minimum TTC value (TTC_{min}) indicates how imminent an actual collision has been during the process of approaching each other.

On average 58% of the 644 encounters between two road users that were analysed in this study had a collision course. For those a TTC curve could be computed and a TTC_{min} determined. At the yield intersection the proportion of encounters with a collision course is about the same for car-car and car-bicyclist encounters, whereas at the general rule intersection the car-bicyclist encounters less frequently result in a collision course, see Fig. 10.

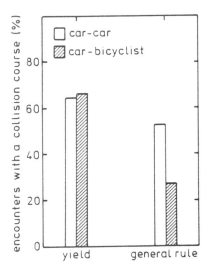

Figure 10.   Proportions of encounters with a collision course by type of encounter and of priority regulation.

The mean TTC_{min} of all encounters having a collision course at both intersections is 3.05 s with a standard deviation of 1.21 s. Because of the skewness of the distribution of TTC_{min}, see Fig. 11, the median TTC_{min} is lower, viz. 2.64 s. The 15th percentile value is 1.94 s. Six out of the 373 encounters with a collision course (1.6%) display a TTC_{min} less than 1.5 s, being only 0.9% of all encounters that were selected (n=644).
At the yield intersection the approach behaviour of car drivers from the minor road seems to be more related to the road environment than to the other road user, since encounters with a party from the right have a higher TTC_{min} than encounters with parties from the left (mean TTC_{min} = 3.3 vs. 2.9 s, respectively), whereas the data on TTI did not significantly differ with respect to the direction the other party is coming from. At the general rule intersection, however, both TTC_{min} and TTI_{min} depend on whether the minor road car has the right of way or not; when its driver has the right of way the mean TTC_{min} is 2.7 s vs. 3.4 s when he has to give way.

4. DISCUSSION AND CONCLUSIONS

The emphasis of this study was on the analysis of what may be considered to be normal road user behaviour in approaching and negotiating intersections, with the aim of developing sound criteria to distinguish between normal and critical behaviour. The use of time related measures is advantageous since they combine several aspects of the dynamic interactive process between road users or between road user and environment into a single variable. Another advantage may be that a description in time makes road user behaviour at the different levels of the driving task directly comparable. In negotiating an intersection road users have to consider potential interactions with other road users. How they deal with them, depends greatly on the type of priority regulation that applies at a given intersection.

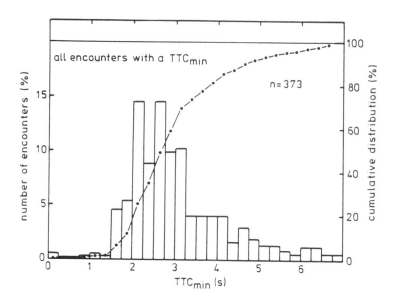

Figure 11. TTC_{min} distribution of all encounters that display a collision course at both intersections. The scale on the right relates to the cumulative distribution.

Two time measures were examined in more detail, viz. Time-To-
Intersection (TTI) and Time-To-Collision (TTC). The TTI measure is related
to the road itself and enables a direct comparison of approach behaviour with
and without other traffic involved. TTC directly relates to another road user
and describes interaction behaviour during the approach process.
    The analysis of drivers' behaviour in terms of TTI gives rather
consistent results. At a yield intersection, for example, minor road car drivers
start braking (defined as the moment the deceleration level exceeds a value of
-1 m/s^2) at a rather constant TTI of about 3 s away from the intersection,
independently of the type of manoeuvre, type of party on the main road, the
direction the party is coming from, or approach speed. The decision to start
braking is made a little earlier. Based on studies on reaction times for braking
(i.e. Johansson and Rumar, 1971; Richter and Hyman, 1974; Glencross and
Anderson, 1976) it takes about 0.6 s for releasing the throttle and applying the
brakes. Taking account of a mechanical response time of the brake system of
about 0.4 s (Malaterre et al., 1987) the decision to start braking is made at
about a TTI of 4 s away from the intersection. Interestingly, at intersections
with traffic signals about 50% of the drivers who are 4 s away from the stop-
line at the onset of yellow, decide to stop (Van der Horst and Wilmink, 1986).
    The control of braking appears to be also rather consistently conducted,
in that, if a minimum TTI (TTI_{min}) occurs before actually entering the
intersection, this minimum is almost always higher than 1.5 s. The occurrence
of a TTI_{min} indicates that the driver has enough time available to come to a

successful stop, if necessary, and the proportion of approaches with a TTI_{min} may reflect the willingness of a road user to stop. At a yield intersection this type of behaviour is more frequently displayed than at the general rule intersection, and is independent of the type or approach direction of the party on the main road. At the general rule intersection, both the proportion of approaches with a TTI_{min} and the mean TTI_{min} values properly reflect whether one has the right of way or not. Head movement patterns also appear to be typical for the type of priority regulation that is in force. Contrary to the findings at the yield intersection where motorists preponderantly start looking to the left, at the general rule intersection the presence of a party coming from the right directs attention to and increases the number of first head movements in the direction of the other party. On average the first head movement occurs at a TTI of about 2.5 s away from the intersection.

To conclude different approach strategies can be distinguished, dependent on the type of intersection one is approaching. This is in line with the ideas of, among others, Hale et al. (1988, 1989), that at a given type of intersection a specific programme of behaviour is started, having its own unique production rules. Selection of the wrong programme may result in severe errors in operation and, consequently, may contribute to the causation of accidents.

If in an interaction between road users a collision course is present (without changing speed or heading a collision will occur), a TTC curve can be computed. In particular, TTC_{min} indicates how imminent an actual collision has been during the process of approaching each other. In this study all encounters between a car from the minor road and a car or bicyclist on the major road occurring during a given period were analysed and represent for the greater part just normal encounters. These encounters hardly ever display a TTC_{min} less than 1.5 s. Two calibration studies on Traffic Conflicts Techniques, i.e. Malmö (Grayson (ed), 1984) and Trautenfels (Risser and Tamme, 1987; Van der Horst, 1990), provide us with both objective and subjective data on critical encounters. About 50% of the interactions scored as conflicts by observer teams in the field appear to have a TTC_{min} less than 1.5 s. Moreover, TTC_{min} is a major factor in explaining the subjective severity of conflicts, although the relationship between TTC_{min} and conflict severity scores is not unambiguous; severe conflicts have a low TTC_{min}, but not all conflicts with a low TTC_{min} are regarded as severe. An analysis of incidents showing a close proximity between the road users, but not scored by any of the observer teams (Van der Horst, 1990) suggests that both the detection and evaluation of critical encounters depends on the type of road users involved. This is in agreement with the finding by Grayson (1984) that conflict type also relates to conflict severity. Moreover, human observers in the field experience detection problems (both 'misses' and 'false alarms') on the border line between slight conflicts and encounters. But once a more serious conflict is detected, the evaluation of severity is rather consistently conducted. A recent example of the use of TTC as a severity measure of near misses in road traffic is given by Brown (this volume).

Based on these results, it is concluded that the TTC_{min} measure is an important variable in discriminating between normal and critical encounters. In line with the results of the TTI analysis of how drivers approach an intersection, a criterion value of 1.5 s appears to be of prime importance. Recently, a first field experiment in which subjects approaching a stationary object with a given speed were instructed to start braking at the latest moment they thought they could stop just in front of the object, confirmed these results (Van der Horst, 1990). Also Time-To-Collision as directly available from the optic flow field seems to be an important cue for the driver in detecting potentially dangerous situations in traffic. The correspondence between the results suggests a fundamental relationship with control and decision-making strategies used by road users in every day traffic. The time-based analysis of road user behaviour may contribute to the further development of road design standards to improve efficiency and safety. For the development of driver support systems based on new technologies, as currently undertaken within the CEC research programme DRIVE, time measures provide detailed information on how drivers operate, what type of information should be provided to the driver, and at what level the system takes over.

REFERENCES

Glencross, D.J. and Anderson, G.A. (1976). Operator Response Factors in the Location and Control of Foot Pedals. Ergonomics, (19)4, 399-408.

Grayson, G.B. (ed.) (1984). The Malmö Study: A calibration of Traffic Conflict Techniques. Report R-84-12, Institute for Road Safety Research SWOV, Leidschendam.

Grayson, G.B. and Hakkert, A.S. (1987). Accident Analysis and Conflict Behaviour. In: Rothengatter, J.A. and Bruin, R.A. de (eds.), Road Users and Traffic Safety, Van Gorcum, Assen, 27-59.

Hale, A.R., Quist, B.W. and Stoop, J. (1988). Errors in routine driving tasks: a model and proposed analysis technique. Ergonomics, (31)4, 631-641.

Hale, A.R., Stoop, J., and Hommels, J. (1990). Human Error Models as Predictors of Accident Scenarios for Designers in Road Transport Systems. Ergonomics, (33) 10/11, 1377-1387.

Hayward, J.Ch. (1972). Near miss determination through use of a scale of danger. Report no. TTSC 7115, The Pennsylvania State University, Pennsylvania.

Horst, A.R.A. van der, and Wilmink, A. (1986). Drivers' decision-making at signalised intersections: an optimisation of the yellow timing. Traffic Engineering & Control, (27)12, 615-622.

Horst, A.R.A. van der, and Godthelp, J. (1989). Measuring Road User Behavior with an Instrumented Car and an Outside-the-Vehicle Video Observation Technique. Transportation Research Record 1213, Transportation Research Board, Washington, D.C., 72-81.

Horst, A.R.A. (1990). A time-based analysis of road user behaviour in normal and critical encounters. PhD Thesis, Delft University of Technology, Delft.

Hydén, Ch. (1987). The development of a method for traffic safety evaluation: The Swedish Traffic Conflicts Technique. Bulletin 70, University of Lund, Lund Institute of Technology, Dept. of Traffic Planning and Engineering, Lund.

Janssen, W.H. and Horst, A.R.A. van der, (1988). Gedrag in voorrangssituaties. Report IZF 1988 C-21, TNO Institute for Perception, Soesterberg.

Johansson, G. and Rumar, K. (1971). Drivers' Brake Reaction Times. Human Factors, (13)1, 23-27.

Malaterre, G., Peytavin, J.F., Jaumier, F. and Kleinmann, A. (1987). L'Estimation des Manoeuvres réalisables en situation d'urgence au volant d'une automobile. Report INRETS 46, Institut National de Recherche sur les Transports et leur Securité, Arcueil-Cedex.

Richter, R.L. and Hyman, W.A. (1974). Research Note: Driver's Brake Reaction Times with Adaptive Controls. Human Factors, (16)1, 87-88.

Risser, R. and Tamme, W. (1987). The Trautenfels Study. Kuratorium für Verkehrssicherheit, Wien.

Chapter 10

USE OF TRAFFIC CONFLICTS FOR NEAR-MISS REPORTING

Gerald R. Brown
Department of Civil Engineering
University of British Columbia
Vancouver, Canada

ABSTRACT

Road traffic accidents are among the most pervasive and serious consequences of the failure of humans to cope with the machine and system infrastructure they are capable of designing. Because of the unreliability of accident reporting and the rare, random nature of accident statistics the use of traffic conflicts, or observed collision avoidance manoeuvres, has been advocated as a means of analyzing road safety. A traffic conflict may be considered a near miss when the collision avoidance manoeuvre reaches some critical level of severity.

What follows is a description of the development and application of the traffic conflicts technique to be used to estimate potential hazard at road intersections, and the suggestion that traffic conflicts may be used for registration and management of road traffic near-misses. The technique embodies a quantitative definition of near-misses as the value of a critical time to collision, as well as a subjective evaluation of driver risk. The application of the technique requires a driver training programme to increase reliability and validity of field measurement. Observing and analyzing traffic conflicts in near-miss traffic situations can help understand the process leading to road accidents and consequently has potential as a means of evaluating road design and traffic control measures.

1.   INTRODUCTION

Road traffic accidents are among the most pervasive, and one of the most serious of our failures to cope with the man/machine environment of contemporary society. The transportation engineer, in conjunction with traffic police, has made major contributions to the design of safe roads by the collection and analysis of road accident statistics. However, it is now recognized that the accident reporting system, and the very nature of accident statistics mitigate against a complete understanding of how road accidents happen, and consequently the most effective management of the problem. Practical and conceptual difficulties with the current accident reporting system are : the poor reliability of accident records (Hall, 1984); the lack of extensive coverage of accidents (Hakkert & Hauer, 1988); the lack of detail which is needed to pinpoint road related causal factors; and the poor statistical characteristics of accident counts because of the random nature and small number of accidents (Hauer, 1986).

These problems have led researchers to suggest the observation and analysis of traffic conflicts to estimate potential hazard. A traffic conflict is conceived by road safety professionals to include all unforeseen collision avoidance manoeuvres, with the most severe conflicts being near-misses. The traffic conflicts technique was conceived by Perkins and Harris (1986) as a method of identifying vehicle crash potential at road intersections. Since its conception the traffic conflicts technique has undergone a series of calibration studies to assess the reliability and universality of the method (Hyden, 1985; Kraaz and Van der Horst, 1985); and has been subjected to validation research directed to finding correlation between traffic conflicts and accidents (Engel, 1985; Glauz et al, 1985; Brown, 1986).

The use of traffic conflicts in road safety studies is an attempt to understand the traffic process leading to accidents by directly observing driver behaviour. This follows from the view that to improve safety we must understand how accidents happen (Oppe, 1985). By the use of field observation of unsafe driver incidents in the traffic stream it is possible to record the road context, the immediate pre-incident conditions and driver response, and the traffic manoeuvre, nature and severity of conflict. A sufficient knowledge of conflicts and their severity, combined with the higher incidence of conflicts vis-a-vis accidents, may lead to the possibility of using traffic conflicts to define the extent of driver risks in a given traffic situation.

In 1984, a project was undertaken at the University of British Columbia to develop an observation procedure for the traffic conflicts technique. Four intersections in Vancouver were studied to validate the procedure, with encouraging results. In 1985, a similar project was undertaken to further validate the procedure, and to investigate the use of the traffic conflicts technique for assessing intersection improvements. Nine intersections were studied to expand the data base for validating the procedure, with six of these nine intersections used to assess intersection improvements. The results indicated that traffic conflicts could be used to diagnose intersection problems, and to measure the positive effects of intersection improvements (Brown, 1986).

In 1987 the procedure was used to analyze the problems of an unsafe, major intersection and this study led to recommended operational and design changes (Brown, 1987). What follows is a report of the procedure which was developed from this work and a summary of the results.

2.   TRAFFIC CONFLICT REGISTRATION

A traffic conflict is based on the notion that collision avoidance by drivers will lead to visual evidence in the form of observable vehicular evasive actions. A traffic conflict involves two road users approaching an intersection at the same point in time and in space, and either one or both road users would have to take some evasive action to avoid a collision. The evasive action (typically some form of braking or swerving) can be identified and rated by observers in the field according to two scales: Time to Collision (TTC); and Risk of Collision (ROC). The time-to-collision measure is the time duration from the beginning of the evasive action of the conflicted vehicle to the point in the roadway at which a collision would have occurred if no evasive action had been taken. The generally accepted upper bound on this measure is a TTC of about 1.5 seconds. See Figure 1 for a graphical description of the TTC concept.

The Risk of Collision measure is the observer's judgement of the chance of a collision occurring irrespective of the Time to Collision observed. This measure incorporates and integrates observations of speed, proximity of vehicles, apparent control of the driving task and environmental conditions such as visibility or a wet road surface.

The TTC and ROC scales used in the Vancouver study were as follows:

Scale Position	TTC	ROC
1	1.6-2.0	"low"
2	1.0-1.5	"moderate"
3	0.0-0.9	"high"

The two scales are combined in some manner to form a composite TTC/ROC severity scale. A simple composite severity scale, made up of 3-point descending scales for TTC and ROC might be:

	Scale Position							
TTC	1	1	2	1	2	3	2	3	3
ROC	1	2	1	3	2	1	3	2	3
COMPOSITE	2		3		4		5		6

Using this composite definition of severity a scale value of 3 or less (1.5 seconds or more on the TTC scale) represents precautionary actions by drivers and would not, under normal circumstances, be considered a traffic conflict. Consequently, a TTC of 1.5 seconds is the threshold value needed before a traffic conflict can be considered to have occurred.

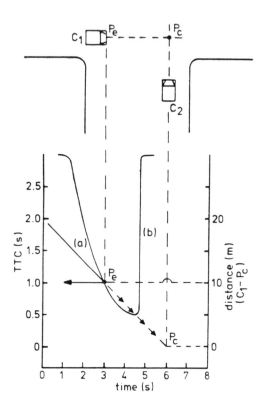

Key: P_e -   point at which evasive action is taken by vehicle C_1.

　　　P_c -   point of potential collision C_1, C_2.

　　　->->   velocity vector of C_1 at instant t_{Pe}.

Figure 1.   Concept of Time to Collision (TTC).
　　　　　(a) as a function of distance
　　　　　(b) as a function of time.

This value might also be used to define a near miss for road traffic situations, and form the basis for road traffic safety management.

3.   OBSERVATION PROCEDURE

Observers are placed in diametrically opposite quadrants of the intersection at least two seconds (at average vehicle speed) from the stop line in order to be able to observe the brake lights of vehicles approaching the intersection. Observers evaluate the context of each conflict incidence and the severity of the evasive manoeuvre taken by the road user. Typically observers look for braking or swerving manoeuvres, the availability of space for a road user to manoeuvre, and the proximity of the two road users during the conflict situation. Essentially, the severity of the traffic conflict depends on the behaviour of the driver, and the (observed) control the driver has over the conflict in relation to the traffic and the road environment. Since an evasive manoeuvre changes the speed and trajectory of a vehicle the TTC can not be measured by mechanical means. For this reason a spatial reference system is needed, requiring a set of reference zones. For each leg, one or more sets of TTC reference zones are established on the approach to the intersection. A minimum of one set of TTC zones are established on the major approaches to the intersection if the width of the intersecting minor approaches is less than 0.5 second times the posted speed (Vp) on the major approaches. Figure 2 shows the reference zones for such an intersection with a posted approach speed of 50 kph.

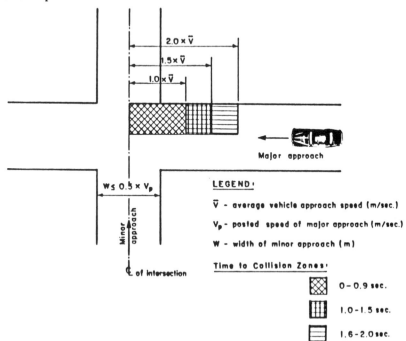

Figure 2.   TTC Reference Zones for a "narrow" Intersection.

Provided that the average speed of the vehicles (V) approaching the intersection does not vary significantly, one set of TTC zones with reference to the
centreline [1] of the intersection would be sufficient for determining the TTC
for most conflicts that occur at the intersection. If a conflict occurs when a
vehicle is travelling at a significantly higher speed than the mean used to
establish the TTC zones, the severity ranking may be increased.

Figure 3 shows an urban intersection with the minor approach wider
than 0.5 second times Vp, and including a pedestrian crossing. For this case
three sets of TTC zones are established: one with reference to the far side
pedestrian cross-walk, one with reference to the centreline, and one with
reference to the near side pedestrian cross-walk.

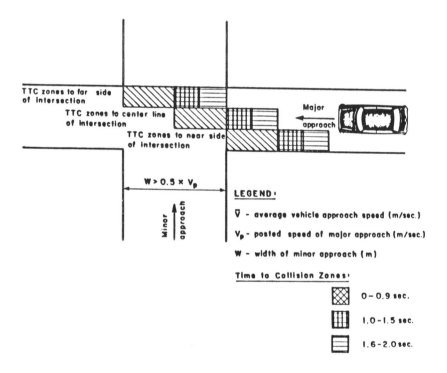

Figure 3.   TTC Reference Zones for a "wide" Intersection.

To improve the specificity of the traffic conflict measures they are invariably
recorded by type of traffic movements being undertaken at the time of the
evasive manoeuvre. Figure 4 shows a typical set of movement categories for
recording traffic conflicts. Note the possibility of ambiguity of conflict,
particularly for the left turn conflicts. For a two observer team (the usual
case) each observer is responsible for recording conflicts on the two
approaches adjacent to him. Double counting may arise when conflicts occur
at the centre of the intersection and/or when both vehicles take evasive action.

---

[1] The use of the centreline gives a practical, though conservative, centroid location for all potential
collision points within the intersection for a typical intersection.

The rule is that the observer best able to see the beginning of the evasive action for any vehicle records the conflict. Double counting is removed in the analysis stage by matching up time of incident occurrence and type of movement from the recording forms.

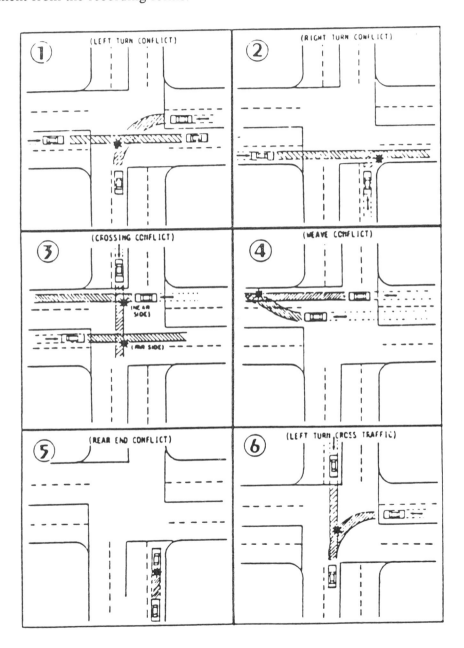

Figure 4.   Traffic Conflict Movement Types.

The length of the observation period can be varied to reflect the traffic conditions of a particular location. Typically, observations are made for at least two weekdays during 8 hours per day, such as 7.30-10.00, 11.30-14.00, and 15.00-18.00. The time of day with the most recorded accidents is also a good indicator for selecting observation periods.

4.   OBSERVER TRAINING

Although the concept is relatively easy to grasp and the observation procedure simple, an untrained observer will have difficulty in consistently identifying and evaluating traffic conflicts. Thus, an ongoing training programme is an inherent component of the procedure.

A practical reason for developing a training programme is to encourage municipalities to adopt the technique to inventory suspected intersections with available staff. Successful implementation of the traffic conflicts procedure in these circumstances requires training of inexperienced observers, and even re-training at regular intervals because of staff turnovers, promotions and such.

The purpose of training is to learn to classify TTC ratings visually without having to perform measurements, and to develop a facility to rate the severity of a conflict. Following brief instruction on the concept of traffic conflict, the training programme involves five successive iterations of field training and video reviews with observer feedback, as illustrated by Figure 5.

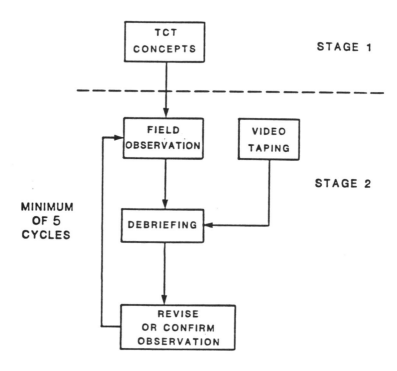

Figure 5.   Observer Training Programme.

After familiarizing the trainees with the identification of the different types of conflict movements, TTC and ROC are discussed and explained. A training film is also shown to the trainees. Familiarizing the trainees with field proce- dure involves observation and recording of traffic conflicts and concurrent video filming for periods of at least two hours at each session.

Figure 6 shows the reliability of traffic conflict recognition by number of training periods for inexperienced observers, and Figure 7 shows the relia- bility in identifying the severity score (TTC and ROC combined). Significant improvements are observed during the first three periods of field training, with reliability of close to 70%. Further training may produce up to 75% reliability but then levels off. It appears that extensive training does not result in substantial improvement beyond 75%. From the results of the effect of a lay-off period it appears that re-training at regular intervals for those observers doing a small amount of conflict recording is also necessary.

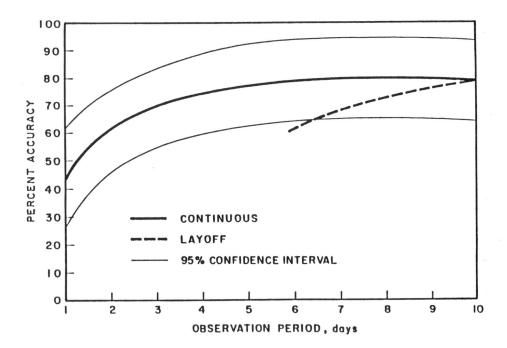

Figure 6. Reliability in Identifying a Traffic Conflict.

5.   AN APPLICATION

The traffic conflicts technique was applied and results contrasted with the accident record for a high volume, unsafe intersection in Vancouver, Canada. The subject intersection shown on Figure 8, is located at the north end of the Burrard Bridge, one of three bridges providing access from the south to the central business district of Vancouver.

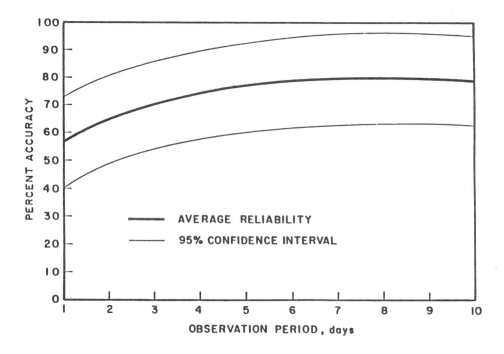

Figure 7.   Reliability in Identifying Traffic Conflict Severity.

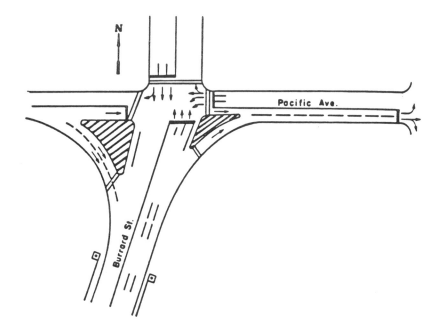

Figure 8.   Application of Traffic Conflicts Technique.

Burrard Street, of six lanes, carries over 3000 vehicles per hour in addition to heavy pedestrian and cycle traffic using the bridge. The traffic record shows a heavy right turn movement from Pacific Avenue eastbound to southbound, and from Burrard Street northbound to eastbound during both morning and afternoon peak periods (1000-1200 vph); and a heavy left turn movement from Pacific Avenue westbound to southbound in the afternoon peak period (584 vph). The afternoon left turn is expedited by a 7 second flashing green arrow at the end of a 19 second solid green fixed time signal interval.

The intersection and approaches recorded 378 traffic crashes in the five years 1982 to 1986. A total of 665 conflicts, disaggregated by movement type were recorded during a 3 day (71 hour) period in July 1987. Sixteen of these were severe conflicts with a composite score of 6. (See Table 1). Comparing the use of conflicts, as opposed to the use of accidents, in evaluating the safety of the intersection gives some interesting results. Firstly, substantial vehicle/pedestrian/cyclist conflicts occurred at the ramp entrance and egress points (59 in all), with 56% of these at the exit from the right turn ramp southbound to the Bridge, and 34% at the entrance to the right turn ramp northbound from the Bridge. In addition there were 34 vehicle/cyclist conflicts with 53% of these at the latter entrance location. These are orders of magnitude greater than the proportional incidence of pedestrians and cyclists in actual crashes (4 injury accidents in each category over the 5 year period). This appears to suggest that pedestrian/vehicle and cyclist/ vehicle conflicts are a major safety problem; while this is not the case when the accident records alone are analyzed. This might be partly explained by the inflated perceived risk to drivers in hitting a pedestrian or cyclist as opposed to hitting another vehicle, and compensating for this added risk by more cautious behaviour.

Secondly, the accident record showed substantial left turn crashes (142, or 49% of major movements) while few conflicts, even serious ones, were recorded for this movement. While many of the crashes associated with left turns occurred at night, and hence outside the period of conflict recording it is also apparent that the evasive action requirement in the definition of traffic conflicts reduces the ability to detect the full risk involved. However, the conflict observations gave good results when the conflicts and accidents were normalized by exposure (i.e. the number of vehicles making the movement; see Table 2).

Thirdly, the conflicts identified a different set of problems within the intersection than did the accident record. Both right turn ramps caused many conflicts, yet accident records indicated rear-end crashes on the ramp approaches were more of a problem in terms of actual collisions. While the pedestrian/cyclist conflicts with vehicles caused few crashes, their effect was presumably transmitted back along the ramp. Thus a cause-effect relationship between the source of driver error and the ultimate results could be observed by the technique.

Table 1. Number of Conflicts by Type of Movement and Severity Score.

Approach (Primary Type of Movement)	6	5	4	3	2	Total
Burrard Southbound (Rear-end)	0*	1	3	1	1	6
	(0)	(0)	(5)	(1)	(1)	(7)
	[0]	[5]	[17]	[20]	[10]	[52]/9.5
Pacific Westbound (Left-turn/Opposing)	0	2	4	6	1	13
	(0)	(1)	(6)	(1)	(0)	(8)
	[1]	[10]	[42]	[32]	[1]	[92]/13.5
Burrard Northbound (Diverge/Xing)	3	4	11	0	0	18
	(0)	(3)	(5)	(0)	(0)	(8)
	[8]	[31]	[65]	[12]	[0]	[116]/16
Pacific East/Southbound (Merge/Xing)	0	3	4	6	0	13
	(0)	(1)	(6)	(7)	(1)	(15)
	[6]	[21]	[101]	[86]	[1]	[221]/16
Pacific Eastbound (Weave)	0	1	8	6	0	15
	(0)	(2)	(8)	(1)	(0)	(11)
	[1]	[15]	[116]	[35]	[17]	[184]/16

*1 = 8-9 a.m.
(1) = 4-5 p.m.
[1] = total conflicts/hrs.

Table 2. Conflict/Exposure Ratios (c/e).

Approach (Primary Type of Movement)	Conflict/Exposure Ratio (c/e)	
	AM	PM
Burrard Southbound (Rear-end)	0.0035	0.0036
Pacific Westbound (Left-turn/Opposing)	0.0158	0.0059
Burrard Northbound (Diverge/Xing)	0.0046	0.0037
Pacific East/Southbound (Merge/Xing)	0.0086	0.0060
Pacific Eastbound (Weave)	0.0078	0.0053
Mean, ($\bar{x}$)	0.0081	0.0049
Standard Deviation, (s)	0.0048	0.0012

The geometric design recommendations arising from the analysis of the traffic conflicts had the effect of reducing the traffic level of service provided by the intersection. It was suggested the number of ramp lanes be reduced, ramp curvature increased, cycle holding areas be removed, and cycle and pedestrian crossings moved away from their former straight line progression. A leading green arrow was also suggested for the permitted left-turn movement instead of the existing lagging green arrow.

6.   DISCUSSION AND CONCLUSION

The traffic conflicts technique may be envisaged as a registration procedure to record near-miss incidents for road traffic. Unlike the accident record the traffic conflict is a direct, observable measure of driver behaviour which may, with some probability, precede a traffic crash.

Systematic observation of traffic conflicts and subsequent analysis of these incidents would appear to have several advantages in diagnosing road safety problems:

(a)   the monitoring of driver behaviour under incipient crash conditions;
(b)   recording of the contextual environment of a hazardous condition, and the driver's response in this context;
(c)   hazard avoidance behaviour may be classified and disaggregated, providing a diagnostic account of pre-incident behaviour;
(d)   incidents can be priorized by level of severity of driver response;
(e)   the road safety engineer has control over critical data items and sample size.

Given the existing reliability and other problems of accident counts it may be that near-misses, as manifested by traffic conflicts, will become an important complementary indicator of roadway hazards.

The technique as now evolving presupposes some level of severity above which a crash is highly likely and below which the crash potential approaches zero probability. The time-to-collision severity index is a quantitative measure of crash probability and driver risk. The present state-of-the-art would suggest that a TTC criterion of 1.5 seconds defines a conflict threshold value, and this value may in turn serve to define a near-miss measure for road safety management. The TTC criterion could perhaps form the basis of a model-driven procedure for near-miss registration and prediction in future work (Van der Horst and Riemersma, 1981).

As a registration procedure for near-miss management the traffic conflicts technique has many positive characteristics. The reporting system is independent of those expected to respond to near-miss avoidance procedures, in that field observers are anonymous, and can record not only near-misses but also certain near-miss pre-conditions and recovery tactics. The development of a data base of near-misses would allow the analyst to monitor the roadway control system for safety efficiency and could result in a fairly subtle and anonymous control mechanism. One disadvantage of autonomously observed traffic conflicts is that no control can be directly extended over attitudes and behaviour imperatives of the driver operator, except perhaps over a long period of time. In addition, the traffic conflicts technique, by observing on-line driver behaviour, gives a more diffuse and complex measure of risk than could be inferred from even very large amounts of accident data. That is, the safety problem is manifested as a multi-dimensional problem. Also while unsafe situations and human error can be measured by traffic conflicts, traffic engineers can only effect the improvement of an unsafe situation by intervening in the operation and design of the physical system, and must to some degree assume a large measure of continuing human error. This perhaps makes the objective of the near-miss management system more difficult than we would desire in more fully controlled transportation systems.

This work on the development of an observational procedure for the traffic conflicts technique, and its application to a high volume intersection has generated two overall conclusions. It is clear that observing and analyzing traffic conflicts at an intersection can help document the root causes of intersection accidents. It is also apparent from the application that the accident record may identify a different set of causal factors than conflicts. At this point in the evolution of the traffic conflicts technique it seems that both observation of on-line behaviour by conflicts and a study of the accident record are necessary for a comprehensive analysis of safety problems. The development work on training observers has led to the belief that it is possible to create a functional team of observers in a relatively short time. However the training period used here is not designed to address all the subtleties of the method, but appears useful for the initiation of a traffic conflict measurement process for management purposes. With further development the technique holds promise as a means of refining the analysis and correction of road safety problems.

REFERENCES

Brown, G.R. (1986). Application of Traffic Conflicts for Intersection Hazards and Improvements. Workshop on Traffic Conflicts and Other Intermediate Measures in Traffic Safety Evaluation, Institute for Transport Sciences, Budapest, Sec. IV.

Brown, G.R. (1987). Traffic Conflicts Analysis for the Intersection of Burrard St. and Pacific Ave. - Vancouver, B.C. Unpublished report prepared for the Insurance Corporation of British Columbia, Vancouver.

Engel, U. (1985). To What Extent Do Conflict Studies Replace Accident Analysis? Validation of Conflict Studies - An International Review. Proceedings of the International Meeting on the Evaluation of Local Traffic Safety Measures, Paris, Vol. 2, pp. 324-343.

Glauz, W.D., Bauer, K.M. & Migletz, D.J. (1985). Expected Traffic Conflict Rates and Their Use in Predicting Accidents. Transportation Research Record 1026, pp. 1-12.

Hall, J.W. (1984). Deficiencies in Accident Record Systems. Transportation Planning and Technology, Vol. 9, pp. 199-208.

Hakkert, S. & Hauer, E. (1988). The Extent and Implications of Incomplete and Inaccurate Road Accident Reporting. Road User Behavior: Theory and Research. Van Gorcum, The Netherlands, pp. 2-11.

Hauer, E. (1986). On the Estimation of the Expected Number of Accidents. Accident Analysis and Prevention, Vol. 10, pp. 1-12.

Horst, A.R.A. van der & Riemersma, J.B.J. (1981). Registration of Traffic Conflicts: Methodology and Practical Implications. TNO Institute for Perception, IZF 1981 C-22, Soesterberg.

Hyden, C. (1985). Conflict Studies and Calibration. Proceedings of the International Meeting on the Evaluation of Local Traffic Safety Measures, Paris, Vol. 2, pp. 366-376.

Kraaz, J.H. & Horst, A.R.A. van der (1985). The Trautenfels Study. Institute for Road Safety Research, The Netherlands.

Oppe, S. (1985). Contribution to Evaluation of Intermediate Variables: Background Paper. Proceedings of the International Meeting on the Evaluation of Local Traffic Safety Measures, Paris, Vol. 2, pp. 317-323.

Perkins, S.R. & Harris, J.I. (1968). Traffic Conflict Characteristics. Accident Potential at Intersections. HRB Record 225, Washington, pp. 35-43.

Chapter 11

ORGANISATIONAL ASPECTS OF NEAR MISS REPORTING

Deborah A. Lucas
Human Reliability Associates Ltd.
Wigan, U.K.

1. INTRODUCTION

The public inquiries into many recent major accidents reported that such disasters could have been prevented if similar (fortunately non fatal) incidents had been registered and acted upon. Disasters such as the Challenger space shuttle, the Hillsborough football stadium, and the Zeebrugge ferry had all been preceded by relevant "near misses". For example, the Sheen inquiry (1987) found that, after the sinking of the Herald of Free Enterprise, previous incidents had occurred where other roll-on, roll-off ferries had left Zeebrugge with their bow doors open. The Presidential report into the Challenger disaster (1986) gave details of numerous incidences of unexpected occurrences of O-ring erosion and "blow-by". The Taylor report (1989) commented on a previous, fortunately non-fatal, crush of football fans at the Hillsborough ground in 1981 at a Cup semi-final. In all these cases (which are not unique in this respect among major disasters) the safety lessons had not been learnt from these earlier precursors. Such cases raise two related questions:

- Why do organisations fail to consider near misses and hence not learn the appropriate lessons?

- How do organisational and management factors influence near miss reporting schemes?

The first question relates to the underlying philosophy relating to the study of near misses; the second issue is more directly concerned with the implementation of a reporting scheme. This chapter will consider both issues briefly.

2.   WHY DO ORGANISATIONS IGNORE "NEAR MISSES"?

2.1.   Central problems of data collection

Reviews of existing data collection systems (see, for example, Lucas, 1987) have highlighted five central problems:

Technical myopia

Most approaches are orientated towards hardware failures rather than human failures. This is despite the known predominance of human performance problems for which figures of between 20% to 80% are cited.

Action orientated

There is often a strong tendency to focus on <u>what</u> happened rather than <u>why</u> the problem occurred. Data is often collated into classes based on the nature of the incident and/or the severity of the consequences. Statistical treatment of incident data only rarely considers causal categories and, as an added problem, there are differing opinions as to the level of causal explanation that is deemed appropriate. This level would appear to be determined by the underlying model of human error causation held by an organisation or its safety department. This model, in turn, is influenced by the perceived role of the organisation in accident and error prevention.

Event focussed

Systems are usually restricted (in practice if not in theory) to looking at individual accidents rather than looking for more general patterns of causes. Hence accident reporting systems are often anecdotal in nature.

Consequence driven

Incidents with serious consequences are recorded and investigated, near misses and potential problems are not often perceived as necessary or worthy of analysis. Even if the advantages of near miss reporting are appreciated, adequate resources of time or personnel are not always available in existing safety departments.

Variable in quality

Reports vary considerably in the quality of description of the facts and in any attempted investigation of underlying causes. Such variations occur both within an organisation and between data collection schemes in different companies. This implies problems with the in-company training of accident investigators and in the lack of systematic methods of incident analysis across all sectors of industry.

Three of these problems are particularly relevant to the issue of near miss reporting. Firstly, the investigation of only those incidents with serious consequences. This problem is fundamentally a question of an organisation holding a reactive management style towards safety which is discussed in detail by Reason, this volume. Secondly, the event focussed nature of current reporting systems which makes accident reporting systems largely anecdotal in nature. The search for patterns of causes is dependent to a great extent on the underlying perception of the causes of accidents and human failures held by an organisation. This model of accidents and errors is a key element of an organisation's "collective memory" and of its prevailing safety culture (Middleton and Edwards, 1990; Westrum, 1988; Lucas, 1991). This issue is discussed in more detail below. Thirdly, the tendency to focus on actions rather than causes which is again related to the underlying model of human error held by an organisation.

2.2.   Models of human error causation

The features of data collection schemes listed above are not only characteristics of the data systems but they also relate to features of the organisation and, in particular, the underlying view of human error causation. Three broad philosophies of how human errors arise in relation to accidents may be distinguished (see Lucas, 1991 for a more comprehensive description).

The traditional safety model

When an accident involving a human failure occurs an investigator holding this approach to human error will typically question the motivation of a person to carry out the system of work safely. The fundamental belief is that errors are caused by a person "not trying hard enough" or "not paying sufficient attention" to the task. This view of human error is closely related to the legal idea of negligence and it is therefore not surprising to find that accident investigations based on this causative model are often concerned with apportioning blame between the worker and management. "Solutions" to the human error problem stemming from this approach will often be limited to disciplinary measures such as dismissal and suspension.

The man-machine interface approach

The second approach to human error is based on a more sophisticated but often passive view of man and human error as a man-machine mismatch. This view maintains that human errors tend to result from a mismatch between the demands of the task, the physical and mental capabilities of the human, and the characteristics of the machine "interface" provided to do the task. This model concentrates on the individual operator and his/her immediate work situation (e.g. the control room). Design changes and the provision of job aids such as procedural support are typical solutions which this view would produce.

Another alternative is the automation of tasks although the inherent dangers of this approach are being increasingly recognised (e.g. Bainbridge, 1987).

The systems-induced error approach

A third approach to human error has emerged over the past decade largely as a result of psychological studies into the role of human failures in accidents. The concept of system-induced error relates to the idea that human failures are caused by certain preconditions in the work context. These preconditions can range from poor procedures, and poor equipment design to unclear allocation of responsibilities, lack of knowledge, and low morale. The important aspect of such preconditions is that many of them can be traced back to management decisions and organisational policies. Indeed, Reason (1989, 1990) argues that poor organisational and management decision making may be "organisational pathogens" in the sense that they may contribute indirectly to accidents. Typical solutions to human error problems using this model of error causation will involve systemic changes in the organisation.

The model of human error causation held by an organisation (through the individuals responsible for safety within that organisation) impacts greatly on the process of accident investigation and the development of control measures. For example, the underlying causes of an incident may not be investigated because of a reactive management attitude to safety which has a remedial, accident driven approach rather than a preventive, problem solving orientation. Such an attitude may result from a "traditional safety culture" in which human error is perceived as being due to the negligence or "carelessness" of an individual. It follows that there will be no attempt to investigate "causes" of the accident other than who did what wrong. Once the "guilty" individual is identified, standard preventive methods (usually designed to increase the motivation of the individual to act safely) are wheeled in. A typical example would be the use of disciplinary actions such as a period of suspension without pay. In such cases, the traditional safety culture and its prevailing "model" of accident causation as being due to a lack of "safety-mindedness" will influence the search for causes and restrict the range of viable alternative remedies.

2.3. Organisational safety cultures and models of human error

Lucas (1991) has proposed that the three models of human error: the traditional safety model, the man-machine mismatch and the system induced error concept may be mapped on to three major types of organisational safety cultures (see figure 1).

Organisational safety culture		Predominant model of human error
Occupational safety management	--->	Traditional safety model
Risk management	--->	Man-machine mismatch
Systemic safety management	--->	System induced error concept

Figure 1: Safety cultures and models of human error

The "occupational safety management" concentrates on the safety of the individual worker and encompasses such concepts as "safety-mindedness". This culture would typically accept the traditional safety philosophy of human error where the motivation of the worker who has erred is questioned. Accident investigations based on this view are usually concerned with apportioning blame between the worker and management. Other safety measures which may be employed with this culture are safety promotions such as posters and "safety blitzes".

The "risk management" culture focuses on the safety of the system and would usually be found in large scale socio-technical systems such as nuclear power and chemical processing. A wide variety of engineering techniques are used to identify hazards and to quantify risks of accidents. It is hypothesised that, where human reliability issues arise, this culture is predominantly associated with the man-machine mismatch view of human error.

Finally, a "systemic safety management" culture may operate when a major organisational disaster has occurred whether this involves loss of life, product liability, or environmental protection. Indeed, Lucas (1991) refers to this culture as being one of "crisis management". It is suggested that this culture may particularly emerge when management liability issues are raised. The culture is concerned with the prevention of disasters through planning and the development of corrective action plans. The systems induced error model would appear to be particularly appropriate for this culture since it involves critical issues of management decision making and the adequacy of organisational systems.

Another tripartite division of organisations according to their safety philosophies has been put forward by Westrum (1988). He has argued that organisations may be labelled: "pathological", "calculative" or "generative" according to their typical responses to safety related incidents. The pathological organisation will tend to deny or suppress information on hazards and may actively circumvent safety regulations. On the other hand, calculative organisations use "by the book" methods but have few contingencies for unforeseen events or exceptions to the rules. The generative organisation accepts that a problem may be global in character and appropriate action is taken to reconsider and reform the operational system. It could well be argued that the pathological and calculative organisations will hold a traditional safety view of error causation. Calculative organisations may also accept the man-machine interface view of error. By their nature the generative organisation must hold a systems-induced concept of human error.

2.4. Near miss reporting and organisational safety cultures

It is proposed that different organisational safety cultures will have an impact on which accidents are investigated and whether near miss reporting is perceived as a valuable use of resources. Typically you would expect the occupational safety culture to be associated with only the recording and investigation of personal injury accidents with serious consequences. With the risk management culture certain classes of near misses may well be investigated. For example, in the nuclear industry, incidents where safety systems have been brought into play are investigated; in aviation, air misses are reported; and on the railways, wrong side signal failures and signals which are driven passed at red are often recorded (see, for example, Taylor and Lucas, this volume). All of these are cases which could result in serious safety consequences. With a systemic safety management culture the organisation may have an even broader perspective in which it is prepared to request reports of perceived "potential problems" from employees as well as investigating near misses and accidents. Such a wide trawl will reveal safety related issues with both immediate and delayed consequences. The latter are particularly interesting since many accident scenarios involve the interplay of failures with latent consequences with more directly active failures.

Westrum's categorisation may also be extended to cover an organisation's approach to data collection (see Lucas, 1989). In terms of accident reporting and incident analysis, the pathological organisation would probably not collect data on hazards and accidents. The calculative organisation would collect such information but would concentrate on applying a traditional model of human error causation to any accident. The key element of this model is that the investigation process is used predominantly to attribute blame. The systems-induced error concept may well be held by generative organisations whose accident investigation process should be more thorough, consisting of a causal back-tracking through immediate operational causes to more distant management attributable "pathogens". It is also to be expected that such organisations will also perceive the value of near miss reporting as a preventive safety strategy to identify the potential for system-induced errors

2.5. Overview

To summarise, it is hypothesised that organisational and management factors will determine the type of accident reporting and analysis which is carried out in any company. In particular, an organisation's safety culture and its related view of the causes of human errors will influence whether near miss reporting systems are used to identify potential problem areas. It follows that a change in the underlying safety philosophy of an organisation may be needed before the benefits of near miss reporting systems are appreciated.

Having outlined those factors which probably influence the general view of safety within an organisation, we are now ready to look at the specific influences on accident and near miss reporting schemes.

3.   ORGANISATIONAL FACTORS IN THE IMPLEMENTATION OF REPORTING SCHEMES

3.1. Design and implementation of reporting schemes

When designing and implementing any accident or near miss reporting system which does not involve the automatic registration of events (but rather voluntary or mandatory reporting by staff) there are a variety of issues to be resolved. Lucas (1987) identified 5 general areas which contribute significantly to a data collection system's success or failure. The first three of these relate predominantly to design issues whilst the remaining two are concerned with organisational and management factors affecting the implementation of reporting schemes. The 5 areas were as follows:

The nature of the information collected

The major factor here is whether the scheme collects mainly descriptive reports (who, what, where, when) or if it additionally covers the causal nature of an error (why). Other factors are: whether near misses are collected, and whether reports consist of written descriptions of the event or text supplemented by answers to specific questions.

The use of information in the database

This factor covers three key aspects. Firstly, whether a particular system provides regular and appropriate feedback to all levels of personnel. To a certain degree this will depend on the second factor of how easy it is to generate summary statistics and pertinent examples from the database. The third factor is whether specific error reduction strategies are generated and implemented by management.

The level of help provided to collect and analyse the data

This item covers the provision of analyst aids in the form of interview questions, decision trees, flowcharts, computer software, etc.

The nature of the organisation of the scheme

Such factors as whether a system is plant-based (localised) or organised centrally, and whether reporting of events is mandatory or voluntary are covered by this item. Additional factors will include: whether the scheme is paper-based or computerised and who is involved in data collection and incident analysis. In general, it does seem that a plant-based computerised system has distinct advantages over more cumbersome centrally-organised schemes.

Whether the scheme is acceptable to all personnel

In this vital area there are at least three key issues. Firstly, the system should have a spirit of co-operation and a feeling of "shared ownership" as opposed to a "Big Brother is watching you" syndrome. Secondly, data should be gathered by a plant-based coordinator who is known to the personnel rather than by an unknown outsider. Thirdly, all plant personnel should receive some introductory training on the purpose of the scheme and the nature of human error. Other aspects which will influence the acceptability of a reporting scheme include: the use of regular and appropriate feedback to personnel, and a system which aims to produce effective solutions to problems.

It can be seen that the traditional view of human error (and the associated tendency to attach blame to individuals who have caused a safety related incident) is incompatible with many of these recommendations for implementing a data collection system on human failures. The safety culture and the model of human error causation held by an organisation therefore affects the implementation of such systems as well as influencing the content and use of the schemes.

3.2. Anonymity, forgiveness and feedback

Three factors under direct management control are vital for the success of any accident and near miss reporting scheme. These factors are: anonymity, forgiveness and feedback. All three aspects influence the acceptability of an accident and near miss reporting system by plant personnel. To illustrate the effect these aspects may have an extract from the report into the Challenger space shuttle disaster is reproduced below.

"Accidental Damage Reporting. While not specifically related to the Challenger accident, a serious problem was identified during interviews of technicians who work on the Orbiter. It had been their understanding at one time that employees would not be disciplined for accidental damage done to the Orbiter, provided the damage was fully reported when it occurred. It was their opinion that this forgiveness policy was no longer being followed by the Shuttle Processing Contractor. They cited examples of employees being punished after acknowledging they had accidentally caused damage. The technicians said that accidental damage is not consistently reported, when it occurs, because of lack of confidence in management's forgiveness policy and technicians' consequent fear of losing their jobs. This situation has obvious severe implications if left uncorrected."(Report of the Presidential Commission on the Space Shuttle Challenger Accident, 1986, page 194).

Such examples illustrate the fundamental need to provide guarantees of anonymity and freedom from prosecution. Once again such guarantees will not be forthcoming in organisations which hold a traditional view of human error. Successful voluntary near miss reporting systems such as the Confidential Human Factors Incident Reporting Programme (CHIRP) run by the UK's RAF's Institute of Aviation Medicine rely on the guarantee of freedom from prosecution to build up their databases.

The third factor, feedback, is also a vital component of voluntary reporting near miss systems. If personnel are to continue providing information they must see the results of their input ideally in the form of implemented error control strategies. A publication which attempts to publicise any insights gained from such a reporting scheme will show all levels of plant personnel that the system is not a "black box" but has a useful purpose. One example of an incident reporting system with an effective feedback channel is the USA's Institute of Nuclear Power Operation's Human Performance Evaluation Scheme (HPES). Here a newsletter "Lifted Leads" is used to publicise anonymous reports of incidents together with any error control strategies implemented. The newsletter is circulated to all plants participating in the HPES programme. In addition, humorous posters have been developed from certain reported incidents and these are also circulated around plants.

4.   CONCLUSIONS

The purpose of this chapter has been to explore some of the organisational and management factors which will influence the success or failure of a near miss reporting system. The major point which has recurred throughout is that an organisational safety culture rooted in a traditional view of human error as due to "carelessness" is fundamentally incompatible with the setting up and effective use of a near miss management system. From a practical perspective, not only will management holding such a view probably fail to investigate the underlying causes of errors but the desire to apportion blame will prevent realisation of the ideals of anonymity and forgiveness. Without such guarantees any near miss management system will quickly fall into disuse.

To set up and maintain an effective near miss reporting system a man-machine interface or system-induced error view of human error is a prerequisite. With either of these models of error causation the investigation of the incident will have more depth and the error control strategies will be more generic. Guarantees of anonymity and forgiveness also sit more easily with these approaches to error.

The lesson from this chapter is that any organisation which is thinking of having a near miss management system must look carefully at its underlying safety culture. If the culture and the related model of human error causation is predominantly one of the traditional safety culture then an initial training programme focussing especially on plant management is probably needed to change attitudes in advance of setting up the incident reporting system. Only with an alternative view of human error will a near miss system prove both beneficial to management and acceptable to users and plant operators.

REFERENCES

Bainbridge, L. (1987) The ironies of automation. In: J. Rasmussen, K. Duncan and J. Leplat (eds.) New Technology and Human Error. London: Wiley.

Lucas, D.A. (1987) Human performance data collection in industrial systems. Human Reliability in Nuclear Power. IBC Technical Services.

Lucas, D.A. (1989) Collecting data on human performance: Going beyond the "what" to get at the "why". Proceedings of a workshop on Human Factors Engineering: A Task Oriented Approach. ESTEC, Noordwijk, Netherlands.

Lucas, D.A. (1991) "Wise men learn by others harms, fools by their own: Organisational barriers to learning the lessons from major accidents. Interdisciplinary Science Reviews, in press.

Middleton, D. and Edwards, D.(1990) (eds.) Collective Remembering. London: Sage.

MV Herald of Free Enterprise. Report of Court No 8074 Formal Investigation. (1987) London: Department of Transport.

Reason, J.T. (1989) The contribution of latent human failures to the breakdown of complex systems. Paper presented at the Royal Society Discussion Meeting on Human Factors in High-Risk Situations, London 28-29 June.

Reason, J.T. (1990) Human Error. New York: Cambridge University Press.

Report of the Presidential Commission on the Space Shuttle Challenger Accident. (1986) Washington DC: Government Printing Agency.

Taylor, P. (1989) The Hillsborough Stadium Disaster. Interim report. London: HMSO.

Westrum, R. (1988) Organisational and inter-organisational thought. Paper given to the World Bank Workshop on Safety Control and Risk Management, 18-20 October, Washington DC.

Chapter 12

CONCLUSIONS

1. OBJECTIVES OF THE CHAPTER

It is always difficult to draw together the threads of a multi-contributor book. The choice of conclusions inevitably reflects the biases of the editors and the interpretations they put on what has been written by the other contributors. We have tried to overcome this bias by writing this concluding chapter jointly and including in it, within a framework, all the features which have struck us either individually or collectively. The reader can add his or her own lessons to the pot and stir at will.

What we have tried to do is to review where we are in the design and use of information systems based on near misses and related incidents; what lessons have we learned from the theoretical discussions and the case studies presented here? what needs to be done in the future to progress the field further? The chapter is therefore organised into three parts:

- issues relating to the *design* of safety management systems and the part played in them by near miss information systems. This section is related to the "ideal" system, or more appropriately to the different "ideals" for different purposes.
- issues related to the *introduction, maintenance and management* of the near miss information system so that something approaching the ideal is realised in practice.
- the many loose ends which suggest *further work*, particularly in the area of validation of the approaches suggested.

2. NEAR MISS INFORMATION WITHIN TOTAL SAFETY MANAGEMENT

Many of the chapters in this book have stressed the pre-eminent importance of defining the purpose of a near miss information system before any decisions are made about its design and introduction.

This is the principle that: "if you do not know what you are designing a wheel for, it will probably turn out square." The answer to the question about purpose may vary considerably depending upon the organisation or system which has to be managed. To structure the possible answers it is necessary to set the question in the framework of a model of safety management activities. Hale (Hale, 1985; Oortman Gerlings & Hale, 1991) considers safety management as the detection and solution of problems within a system. This process goes on at two levels:
- the detection of problems new to the system so that preventive design changes can be made and a safety management system can be set up to keep the remaining risks under control.
- the detection and correction of failures in the safety management system, where known hazards are not successfully being kept under control.
The steps in the problem solving cycle for the two levels of activity are similar; see figure 1 for a generic description.

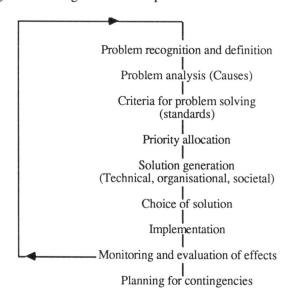

Problem recognition and definition
|
Problem analysis (Causes)
|
Criteria for problem solving
(standards)
|
Priority allocation
|
Solution generation
(Technical, organisational, societal)
|
Choice of solution
|
Implementation
|
Monitoring and evaluation of effects
|
Planning for contingencies

Figure 1.   Safety management as problem solving.

Using this model the contribution of information systems based on near misses and related incidents can be located at three points in the cycle (see also Chapter 1):
- problem analysis leading to priority allocation and the generation of appropriate solutions. How can the system go wrong in ways which are not yet known? This use is aimed at *system modelling*.
- *monitoring* of the effectiveness of the current safety management system. Which existing prevention measures are not working as well as planned?
- problem recognition, the proof that a new system is vulnerable or that an old one is still vulnerable despite current attempts to make it safe and hence

that vigilance in the management of safety should not be relaxed. We call this use *"motivation"* because it serves to activate people in the organisation to be and remain alert.

All three purposes are related clearly to actions that the system managers need to and must take; respectively introducing new preventive measures, modifying or enforcing the use of existing measures and re-emphasising the importance of vigilance in relation to current prevention measures. The type of data required and the way in which it needs to be handled to satisfy each purpose can be different.

2.1. System Modelling

This purpose is relevant if we do not know how and why the socio-technical system is breaking down and leading to accidents or at least, that we do not know in sufficient detail to be able to plan or choose the relevant prevention measures.

This purpose received the greatest emphasis in the chapters presented here. It is the purpose which interests the academic researcher the most, because it presents the most fundamental challenge to produce new and universal knowledge. It is not surprising that the emphasis in these chapters (e.g. Reason, Van der Schaaf in Chapter 6, Masson, Taylor & Lucas) is on getting a deeper understanding of the human factor in the system. This is the element which is still far less understood than the hardware of the system. This is particularly so in the human activities which are the least regimented, e.g. car driving, or where current human behaviour models are the weakest, e.g. in knowledge based behaviour under emergency situations.Where behaviour is simpler or more constrained by the system rules, e.g. in routine work in industry, system modelling at the level of individual behaviour is more advanced; there the level of modelling which is still in its infancy is the organisational control of hazards.

The implications for a near miss information system of an emphasis on modelling are the following:
- the system is mainly interested in *"unusual" incidents*, which cannot be predicted or understood by existing models.
- the system is more interested in *qualitative data* than in quantitative. The first time that a given type of incident occurs it demands an expansion of the system model to explain it. The second time it occurs is little more than a confirmation that what is happening is not a freak event. After that the interest passes to the question whether the probability and consequences are great enough to warrant action (a monitoring function).
- the system demands *deep and rich information* about the incidents in order to give enough data on which to perform the modelling. Data will be needed on all three aspects of the incident; the hardware, the behaviour and the procedures and organisation. It is the interaction of the three which is most often not sufficiently modelled.

2.2. System Monitoring

This purpose is relevant if we already have a safety management system with preventive measures designed to combat given types of incident. In this case the interest is to detect whether those types of incident are still occurring in sufficient numbers to indicate that specific preventive measures are not working.

The implications for the near miss information system of this purpose are:
- it must be driven by a *model* which links type of incident with type of preventive measure.
- it is most interested in *quantitative data* ; *that* the incident happened and not how (since this is already known). The system is interested most of all at the organisational level, in how and why the existing safety management system allowed this known problem to recur.

2.3. Motivation

An issue underlying a number of the case studies, which also emerged in the discussions at the CEC workshop, is that of the *credibility of risks*. Safety management tries to predict and eliminate or control all significant risks in a system. The number of imaginable ways that things can go wrong is usually much greater than the number of ways that a system actually will go wrong (or has gone wrong) in its life cycle. Those managing the system are therefore always faced with deciding which risks are credible and should lead to action and which ones can be ignored (or accepted). Accidents have the notorious habit of concentrating the mind of the managers on preventing that particular failure happening again, sometimes to the detriment of other necessary actions. The absence of overt accidents can also have a strong effect on the minds of the managers, leading them to think that all is perfectly well.

Both of these problems can be tackled to an extent with near miss reporting systems which indicate;
- that there are other measurable deviations happening even in systems with low accident rates. These provide information on which future watchfulness can be planned.
- that besides the accidents which have happened in the less well controlled systems a range of other potential accident types are lurking just below the surface.

The data fulfils in this sense a motivational role to keep the level and focus of alertness adjusted. Akin to this is the need for *training* for new operators and managers who have no direct experience of the things which can go wrong in the system. For them near miss and accident data can be a substitute which convinces them that particular sorts of accident are indeed credible.

In order to fulfil this purpose of the near miss information system the data needs to satisfy the following criteria:

- be sufficiently *rich and descriptive* to convince the manager of the reality and credibility of the incident.
- have an emphasis on the way in which the incident could have been *prevented* (otherwise the lesson learned will be that accidents are such odd events that they can never be controlled).

2.4. Comparing purposes

In any practical situation the near miss information system will have to fulfil a *combination of the three purposes* outlined. A vital stage in designing any system is to decide how that balance lies. From the summary given it is clear that the biggest contrast is between the modelling and monitoring purposes, one requiring rich qualitative data about unusual incidents, the other model-driven quantitative data about known types of incident.

A similar difference of emphasis between modelling and monitoring may reflect the different viewpoints of the academic researcher and the practical system manager. The researcher is often frustrated by the shallow-ness of the information collected by most routine incident reporting systems, whether industrial or transport. They do not begin to help in model building. The reaction may be to try to make the routine collection richer and deeper. This in turn risks overloading the organisation with time-consuming work which it sees as useless in relation to most of the incidents which occur. It seems logical that an information system required to fulfil both modelling and monitoring purposes will need to have a *hierarchical structure*, which operates at a shallow monitoring level for the bulk of known problems and only collects deeper data on the unusual events, or on the types of incident (identi-fied out of that shallow pattern recognition) which despite the current pre-cautions still keep occurring, and so need modelling work.

2.5. Near misses and accidents

All three purposes can be fulfilled by accident data as well as near miss data. The advantage of the latter is that they are more frequent and may be collectible with less emotional biasing than the former. In particular the motivational purpose in systems which have highly effective safety management systems, and which have already succeeded in eliminating most full-blown accidents, may only be fulfilled with near miss data. Near miss data also has the potential advantage that it can place emphasis on the way in which the incident was *recovered* and so did not become an accident. Such an emphasis gives a positive side to the information system, which may be psychologically important. It can also help to concentrate the attention onto ways of strengthening the recovery systems in the future.

These advantages of near miss over accident data came out clearly in a number of the chapters (e.g. the Introduction and all case studies). The limitations and possible *disadvantages of near miss data* received less attention (see however Chapter 2 for Reason's critique, and Chapter 3 for a response). The very advantage of near misses of concentrating on recovery has a negative side. Accidents are by definition near misses which failed to be recovered in time. The events which the safety management system wishes to prevent are therefore just those events which will never, by definition, occur in the near miss information system. It is vital, therefore to be clear just *what near misses have in common with accidents and what not.* Van der Horst's chapter tackles this question directly and arrives at a pragmatic, but clearly measurable definition of interesting traffic conflicts which relates them to accidents. Taylor and Lucas have a definition based on one clearly measurable event (passing the signal at danger) related to one type of accident risk. In both cases the near misses are clear enough precursors of accidents to serve as reasonable predictors not only of the qualitative type, but also of the quantitative risk of accidents. In other cases (e.g. Van der Schaaf in Chapter 6, Hale et al), which are trying to collect near misses in relation to many different actions and risks, the definition of what to report as a near miss has to be left very much up to the discretion of those reporting. Such reliance on subjective definitions of what constitutes a near miss is not a problem for a system interested in modelling. The information system is interested in incidents from which it can learn; this is similar enough to the motivation which can be assumed to lie behind spontaneous reporting of incidents - that they seem to the reporter to be worrying, surprising and worth looking at further. If the number of reported near misses is great enough there is a reasonable chance that significant new accident types will be discovered. Where the monitoring function is dominant, there is such a built-in quantitative unreliability in such a reporting system that it will cause problems. Such monitoring systems need much more clearly defined lower reporting limits, such as "injury needing treatment" or "signal passed at danger".

What is common between the near miss and the accident is the steps from deviation up to the point where, in the former case, the recovery action was initiated (or where other more or less "chance" events stopped the development towards the accident). *Near miss data collection must therefore concentrate on these early steps if it is to be of value in modelling the accident-causation process.* Systematic data on recovery can also be of value, but only if it is carefully examined to see what evidence it reveals of latent flaws in those recovery mechanisms; what went right, but could very easily have gone wrong? For monitoring purposes the recovery data from near miss events can give valuable information about the prevention measures which are the most often challenged and which therefore stand "in the front line" of prevention.

The workshop also clearly defined the need to draw a lower bound to what was called a near miss. If the system strays too much into the realm of collection of all errors it becomes too cumbersome. It will also lose the credibility of those who have to run it and report the incidents. They may no longer believe in the learning value of the things which they report.The *definition of a near miss* as "a deviation which has clearly significant potential consequences" therefore seems a good pragmatic one.

3.   INTRODUCTION AND MAINTENANCE OF INCIDENT REPORTING SYSTEMS

The general framework of seven steps (figure 2) proposed by Van der Schaaf in Chapter 3 met with broad acceptance as the way to think about and plan the design of the system.

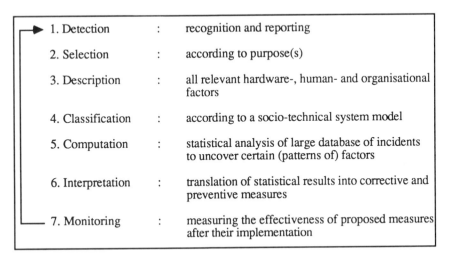

1. Detection	:	recognition and reporting
2. Selection	:	according to purpose(s)
3. Description	:	all relevant hardware-, human- and organisational factors
4. Classification	:	according to a socio-technical system model
5. Computation	:	statistical analysis of large database of incidents to uncover certain (patterns of) factors
6. Interpretation	:	translation of statistical results into corrective and preventive measures
7. Monitoring	:	measuring the effectiveness of proposed measures after their implementation

Figure 2: The seven modules of the basic NMMS design framework (see also Chapter 3).

The previous section has indicated that such a model needs to be equipped with *more iterative loops* than the single feedback loop in figure 2. "Classification" may follow "description" in a modelling system, but for a monitoring system the classification almost replaces the description and is even built into the selection of incidents. For a modelling system the selection of incidents to study in depth will change over time as one type of incident becomes better understood and modelled, and so can be passed over to the monitoring system.

One of the major lessons from the workshop was the paramount importance of the organisational in-bedding of the incident reporting system. Many of the chapters have emphasised how easily the objectives of a system can be frustrated by problems in introducing it into an organisation, or by changes in the way in which management sees or uses it. This means that the seven steps need to be placed in a broader framework defining the life cycle of such an information system (figure 3).

3.1. Objectives and design

The most important driving force of any reporting system is the *motivation* of those who must do the work in it, i.e. those who see and report the near misses. We have already made, on theoretical grounds, the distinction between system modelling and monitoring as possible objectives of the system.

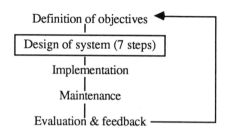

Figure 3.   Life cycle of the NMMS.

In considering motivation at all stages in the life cycle of the information system this distinction plays an even greater role.

Traditionally accident reporting systems have been associated in the minds of many with judicial proceedings, the allocation of blame and the taking of disciplinary action (see the chapter by Lucas). They are seen as repressive or corrective instruments of management, and as such have always suffered from problems of underreporting and hiding of information, particularly where that may be related to potentially blameworthy actions of individuals. This picture has the closest similarity to the monitoring purpose described above, but then with a strong flavour of *monitoring the individual rather than the safety system as a whole*. If near miss information systems are set up to have a monitoring function this point needs to be confronted and solved. The chapter by Ives indicates what happens to a system which is used to monitor behaviour in such a negative context; the reporting dries up and the system collapses. This will be particularly so where there is no objective way of detecting the near misses which should have been reported. Even where such a method does exist, the feeling that the system is being used as a "big brother" will cause resentment and a negative attitude to the truthfulness and value of the data. The only path open is to stress that the data is monitoring the system and not the individuals in it; e.g. that an incident arising from not following laid-down safety rules can just as (or even more) easily be a problem with the rules rather than with the individual breaking them. It can be difficult to convince people that such a statement of policy is truly meant. They may remain suspicious until it has been proved in practice. Such *problems of trust* may even be significant enough to decide that system monitoring should be only done with an accident reporting system, leaving the near miss system with a purely modelling role.

A system aimed at modelling can get over this negative image by stressing the creative, professional and scientific purpose of the data as a *learning device*. This will only be believed if there are very clear guarantees that reporting of a near miss will never result in disciplinary action aimed at those involved (see Chapters 6 and 11). It will also only be believed if the operators themselves are closely involved in the design of the reporting system, the analysis of the data and the decisions about action to be taken. It is clear from studies such as Van der Schaaf's in Chapter 6 that, if such criteria are met, the information system can be a powerful stimulus to involvement of operators in the active improvement of safety.

3.2. Implementation

From the chapters of Hale et al, Van der Horst and Van der Schaaf (Chapter 6) the importance of *training* in the implementation phase is clear. Those who are reporting and those who are classifying the incidents have a profound influence on the value of the data. It is vital that they have a clear model of how accidents occur (e.g. the deviation model referred to on numerous occasions), of what factors are relevant to be recorded and of the objectives of the reporting system.
- Managers must be trained to use accidents not in terms of guilt and blame, but in terms of a socio-technical system failure to which they must respond with a system design change.
- Operators must be trained what to report and why it is important.
- Investigators must be given appropriate models of the complexity of causal chains in accidents, leading back to all levels in the organisation and the way it works.

Since many of these people (particularly at operator level) will have relatively unsophisticated ideas about accident causation to start with, this is a significant training burden. The chapter by Hale et al indicates that an interactive computer-based registration system can offer some help, but that there is still some way to go before it can be widely used.

It was also clear from the discussions in the workshop that a common mistake in implementing systems was to see them as packages to be sold on a turn-key basis, rather than as concepts to be adapted to the specific needs of the organisation. *Implementation is therefore a gradual process*, which needs to have feedback loops built into it to arrive at the optimum organisational in-bedding. It is even a question whether such an optimum is ever reached, because the purposes of the system may change over time. This point is developed in the next section.

3.3. Maintenance and evaluation

It is clear from the discussion in section 2 that a near miss system aimed at system modelling is inherently dynamic. As soon as it has collected enough data to make progress with the modelling and decide upon appropriate pre-vention measures, its focus changes and that type of incident moves across to the monitoring objective. This means that the criteria for selecting incidents for deeper analysis will constantly change.

We have stressed the importance of involvement of the reporters in the analysis and interpretation and implementation phases as an incentive to main-tain reporting. A particular problem arises with information systems designed to serve widely separated levels in the hierarchy. If the reports are made at the shop floor/by the driver and the analysis and decision making are done higher up in the organisation, or at a remote headquarters site, a very strong *feed-back loop* to the reporters is needed, consisting of information, encouragement and demonstration of the value of the data generated (see Lucas' chapter). If this is not done, reporting will gradually fade out.

Such a dilemma occurs particularly where incidents occur only rarely and in remotely scattered locations. They then have to be *collected centrally for statistical analysis* (especially for monitoring purposes). If the richness of data, which can only be *generated locally*, is not to be lost, the reporters have to be given some incentive to cooperate. Using the data locally for system modelling and learning and sending it through to a central point for monitoring purposes may be a way out of this problem.

In the chapters by Brown and Van der Horst this problem is side-stepped because the researcher was both reporter and analyst. Conflict observation was used as a one-off tool to investigate a problem which had already been picked up by the use of other data analysis. It is hard to see how it could be adapted to the sort of routine use that is characterised by the industrial or railway studies.

4.   FUTURE DEVELOPMENTS

"Horses for courses"

A unique feature of this book is that it sets out applications of near miss information systems from a range of industrial and transport settings next to each other. This allows some parallels and differences to be emphasised. More such comparisons in the future were felt to be desirable.

The issues which arose from the comparisons were the following:
- different transport and industrial organisations are at a different point in their development with respect to the purpose of near miss reporting. The conventional industries placed a strong emphasis on the need to get into the procedural and organisational determinants. This was also a matter of concern for the high technology industry and rail and air transport organisations represented, but they also placed great emphasis on the value of near miss data in unravelling (modelling) of human behaviour (operators/ drivers/ pilots/ air traffic controllers). The road traffic examples are almost exclusively related to driver/road interactions, with no concern (as yet) for organisational considerations.
- the intervention and control options available in the different systems are also important factors determining the shape of the near miss reporting. In road transport the individual driver cannot easily be involved either as reporter of near misses, or as focus of short term prevention measures. This lies differently in professional transport systems and in industry. In both these cases the range of preventive measures is greater and more strictly enforced; this makes the monitoring function of incident reporting systems more salient.

4.1. The dynamic aspects of reporting systems

We have emphasised in this chapter the dynamic features of near miss reporting systems, and the relationship they must have to other safety management information systems serving the wider purpose of the control of hazards in the given system. This has not been a feature of the studies so far made of reporting systems. They have been seen far more as static systems, which you get right once and then leave to function. We would urge *greater attention to charting the life cycle of such information systems*, so that their managers can be given much more guidance about how to steer and modify them in response to changing needs; what are the signs that a system is becoming obsolete? how can change and continuity be balanced?

In relation to this last point, we acknowledge the value of long term trends in data as an aid to monitoring safety system effectiveness. This demands *consistency in data recording and classification over time*. The higher up the organisation one goes the more such overall data is used. At the top of this scale are the national and European recording systems. We have stressed the essential need for data systems *at the grass roots to be flexible and dynamic* in their focus of recording and interpretation. The integration of such contrasting local and national requirements into a hierarchical system which can meet both needs is a challenge which requires much future work. The solution might be sought in a difference in emphasis at the different levels between near miss and accident data.

We have also stressed the potential links between near miss and other safety information systems. Here too more thought and study needs to go into the development of integrated systems, or into a clearer division between them, if their purposes are shown to be fundamentally incompatible.

4.2. Need for evaluation

Our final call is for more evaluation of the systems which are in place or being introduced. It is striking that the case studies presented concentrate much more on the early steps in figures 2 and 3 than on the monitoring and evaluation. It is often an embarrassing question to ask to the person responsible for an information system: "what decisions can you point to as having been taken as a result of the data from this system?" It may be even more embarrassing, even if the system passes that first test, to ask: "could the same decision have been taken based on information from another and cheaper source?" Many accident reporting systems are based on the apparently self-evident value of recording data, because it is there and because the law requires it. The data may then languish too often in the drawer or computer into which it is loaded. It would be a pity if near miss information systems followed that bad example.

The evaluative information offered in this book is also largely anecdotal and concentrates on whether the organisation accepted or rejected the system as a whole. We need to progress to the next stage of detailed analysis of the value of different types of data for different purposes.

4.3. Beyond near miss reporting

The system modelling function of near miss reporting has been stressed in this chapter. The aim of such modelling is to produce such a good understanding of how systems produce accidents that countermeasures can be designed into them to prevent the accidents before they happen. This utopia has come somewhat closer by with the considerable advances in modelling of the human element in systems in the last decade. The chapter by Masson may be a look into the future at the next step in that process. It describes the building of a software model of a system which can be run to check whether it creates errors and if so what sort of errors. Such simulation models have been a commonplace of hardware design for a number of years now. We are only on the threshold of the developments necessary to extend it to the human and organisational elements in systems. If it realises that potential it will shift the emphasis of near miss reporting from an aid in primary system modelling to a means of evaluating whether the system modelling is accurate enough and to the monitoring role we have also discussed. Before that happens a lot of near misses will need to have flowed through our reporting systems.

REFERENCES

Hale, A.R. (1985). The human paradox in technology & safety. Inaugural lecture. Delft University of Technology.

Oortman Gerlings, P.J. and Hale, A.R. (1991). Certification of safety services in large Dutch industrial companies. Safety Science, 14 (in press).

Index

f indicates that the item extends over the following pages of that chapter